MW01253083

Knowledge Flows, Governance and the Multinational Enterprise

Also by the editors

Volker Mahnke, COMPETENCE, GOVERNANCE AND ENTREPRENEURSHIP (*with N. Foss*)

Volker Mahnke, THE STRATEGY PROCESS (*with M. Venzin and C. Rasner*)

Torben Pedersen, MANAGING CENTRES OF EXCELLENCE (*with U. Holm*)

Knowledge Flows, Governance and the Multinational Enterprise

Frontiers in International Management Research

Edited by

Volker Mahnke

and

Torben Pedersen

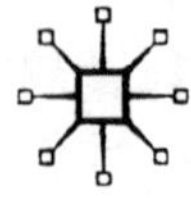

First published 2004 by
PALGRAVE MACMILLAN
Houndmills, Basingstoke, Hampshire RG21 6XS and
175 Fifth Avenue, New York, N.Y. 10010
Companies and representatives throughout the world

PALGRAVE MACMILLAN is the global academic imprint of the Palgrave Macmillan division of St. Martin's Press, LLC and of Palgrave Macmillan Ltd. Macmillan® is a registered trademark in the United States, United Kingdom and other countries. Palgrave is a registered trademark in the European Union and other countries.

ISBN 1–4039–3311–1

This book is printed on paper suitable for recycling and made from fully managed and sustained forest sources.

A catalogue record for this book is available from the British Library.

Library of Congress Cataloging-in-Publication Data
Knowledge flows, governance and the multinational enterprise: frontiers in international management research/edited by Volker Mahnke and Torben Pedersen.
p. cm.
Includes bibliographical references and index.
ISBN 1–4039–3311–1 (cloth)
1. Knowledge management. 2. Technology transfer – Management. 3. Corporate governance. 4. International business enterprises – Management. 5. Business networks. I. Mahnke, Volker. II. Pedersen, Torben.
HD30.2.K63634 2004
658.4′038—dc22 2003058147

10 9 8 7 6 5 4 3 2 1
13 12 11 10 09 08 07 06 05 04

Printed and bound in Great Britain by
Antony Rowe Ltd, Chippenham and Eastbourne

Contents

Part III Governing Internal Knowledge Relations

Part IV Knowledge Governance and Business Development

List of Figures

List of Tables

List of Contributors

Ulf Andersson	Uppsala University, Sweden
Niklas Arvidsson	Stockholm School of Economics, Sweden
Julian Birkinshaw	London Business School, UK
Anders Blomstermo	Stockholm School of Economics, Sweden
John A. Cantwell	Rutgers University, NJ, USA and Reading University, UK
Doren Chadee	University of Auckland, New Zealand
Charles Dhanaraj	Indiana University, IN, USA
Mats Forsgren	Uppsala University, Sweden
Jens Gammelgaard	Copenhagen Business School, Denmark
Anna Gatti	UC Berkeley, CA, USA
Ulf Holm	Uppsala University, Sweden
Robert Johnston	Warwick University, UK
Valerie J. Lindsay	University of Auckland, New Zealand
Mark Lorenzen	Copenhagen Business School, Denmark
Marjorie A. Lyles	Indiana University, IN, USA
Volker Mahnke	Copenhagen Business School, Denmark
Jan Mattsson	Roskilde University, Denmark
Ram Mudambi	Temple University, PA, USA
Margit Osterloh	University of Zürich, Switzerland
Joanne E. Oxley	University of Michigan, MI, USA
Torben Pedersen	Copenhagen Business School, Denmark
D. Deo Sharma	Stockholm School of Economics, Sweden
Benjamin Ståhl	Uppsala University, Sweden
Morten Thanning Vendelø	Copenhagen Business School, Denmark
Antoinette Weibel	University of Zürich, Switzerland

Part I

Fundamental Perspectives on Knowledge Governance

1
Knowledge Governance and Value Creation

Volker Mahnke and Torben Pedersen

Introduction

Since its inception the field of MNC research notices that knowledge is key in the explanation of the existence, boundaries, internal structure, and competitive advantage of the MNC. No doubt, the governance of knowledge flows has remained a central theme in a large body of current academic work concerned with managerial practices in the MNC. The modern MNC is considered to be a 'differentiated network', where knowledge is created in various parts of the MNC and transferred to several interrelated units (Hedlund, 1986; Bartlett and Ghoshal, 1989; Gupta and Govindarajan, 1991, 2000). In this perspective, MNCs are no longer seen as repositories of their national imprint but rather as instruments, through which knowledge is accumulated in various local contexts and transferred horizontally between subsidiaries, shared vertically in headquarters–subsidiary relations, and between the MNC and its environment.

However, despite continuing interest an integrative model on how to govern knowledge in the MNCs has not emerged yet. Little is known about how particular governance arrangements support value creation in the MNC. The objective of this chapter is to contribute to an integrative understanding *of how the governance of internal and external knowledge flows contributes to value creation in the MNC.*

As illustrated in Figure 1.1, we identify key elements crucial in the empirical development of an integrated understanding of knowledge governance in the MNC, including (a) alternative knowledge flows, which are (b) complicated by cognitive and motivational challenges to which (c) governance mechanisms respond in the MNC's attempt to create value. Next we distil from the several contributions to this

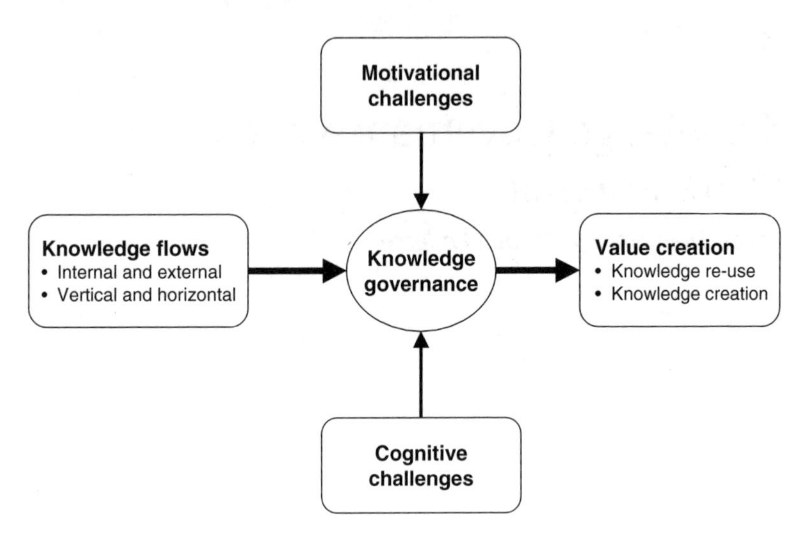

Figure 1.1 Knowledge governance

volume some key findings on the interrelation between these elements to reach conclusions on whether and how certain constellations of knowledge governance are conducive to an MNC's attempt at creating value. Conclusions for further research follow.

Knowledge flows and value creation

We define a knowledge flow as a process (Szulanski, 1996; Argote, 1999, Zahra and George, 2002) that covers several stages starting from identifying knowledge over the actual process of transferring the knowledge between participants to its final utilization by the receiving unit. This definition draws attention to (a) cognitive and motivational challenges in receiver–sender relations, (b) the determinants of costs involved in various process phases and (c) the need for knowledge utilization to create value.

Knowledge gained through participating in knowledge flows yields value in two principal ways, either through knowledge re-use and/or knowledge creation. The former concerns knowledge that one party to an exchange already has and another party needs, for example, when the MNC's headquarters needs to make an informed decision that requires knowledge about local markets or if one subsidiary can benefit from best practices developed in another. Other knowledge flows, by contrast, create value by developing new knowledge through combination of existing knowledge or through the innovative application of existing knowledge to new market contexts.

To create value, participants in a knowledge flow require the ability and the motivation to understand, assimilate and apply knowledge (Cohen and Levinthal, 1990; Lane, Salk and Lyles, 2001). Similar, Zahra and George (2002) summarized representative empirical studies to argue that acquisition, assimilation, transformation and exploitation of knowledge are equally important elements. By implication, parties to a knowledge flow may acquire and assimilate knowledge but might not transform and exploit the knowledge for profit generation. In addition, because their motivation and ability might differ, parties to a knowledge flow might profit unequally from participation. Accordingly, one can argue that knowledge governance must be responsive to the type of knowledge flow as well as to constellations of motivation and ability among participants.

Governance challenges: cognition and motivation

There are two critical challenges to be addressed in the governance of knowledge: the challenge of motivation and the challenge of cognition. The motivational challenge refers to the constellation of interest among parties to a knowledge flow, while the cognitive challenge refers to knowledge and information asymmetries that complicate exchange despite best intentions among participants (e.g. Gupta and Govindarajan, 1991, 2000; Szulanski, 1996; Buckley and Carter, 1999). For example, even if agents in a subsidiary can be motivated to take actions that are incentive-compatible with those of other subsidiaries or the MNC headquarters, there is still no guarantee that they also make optimal (i.e., value-maximizing) choices if cognitive challenges remain. In other words: willingness is not the same as ability to engage in value-creating knowledge flows. Because both motivation and cognition matter for the governance of knowledge flows we shall discuss these associated challenges in turn.

Cognitive challenges

Cognitive challenges have important implications on the costs and benefits of alternative ways to organize knowledge flows in the MNC.

To address cognitive challenges, one can make a distinction between codified and uncodified knowledge (Nonaka and Takeuchi, 1995; Kogut and Zander, 1992; Boisot, 1998). Whereas codified knowledge often results from abstraction and establishing cause–effect relation expressed in written form, uncodified knowledge often results from local experience, is context-dependent and remains often embodied in the MNC's employees. Whatever the directionality and purpose of knowledge flows in the MNC, one can expect that the larger the proportion of uncodified knowledge in the MNC, the more costly knowledge identification and integration will be compared to a knowledge flow situation where knowledge is codified. Thus, managers in the MNC face two choices: either (a) they articulate and codify knowledge to decrease the costs of knowledge transfer, or else (b) rely on costly personal communication channels.

A second dimension of cognitive challenges concerns the distinction between common and partitioned knowledge assets (Bartlett and Ghoshal, 1989; Buckley and Carter, 1999). Knowledge in the MNC might be commonly understood or alternatively highly partitioned, for example because it is developed locally. As Grant (1996) argues, the greater the degree of commonly shared knowledge, the easier knowledge integration becomes. At the same time, the more knowledge is commonly understood the less can be gained through knowledge integration. Thus, while partitioned knowledge resulting from dispersed learning is conducive to knowledge diversity and potential value creation through knowledge combination, integrating partitioned knowledge is likely to be more costly the greater the knowledge gaps and the lower the degrees of commonly shared knowledge among parties to a knowledge flow.

Regarding external knowledge flows, the same dimensions can be used to describe implications for governing knowledge flows between an MNC and its external partners. As the costs of knowledge flows tend to rise with increasing degrees of uncodified knowledge and knowledge-partitioning, learning from external partners will be the more costly the less knowledge is commonly shared and the less knowledge is codified (e.g. Hamel, 1991; Lyles and Salk, 1996). A further complication arises to the extent that knowledge of partners is less codified than knowledge of the MNC: the risks of knowledge expropriation increase while learning possibilities remain limited (e.g. Hamel, 1991). By implication, codifying knowledge is a double-edged sword: internal search and communication costs decrease but the risk of external imitation increases. In addition, this risk is particularly pronounced if knowledge gaps between external partners and a particular subsidiary are smaller than the knowledge gaps between the subsidiary and other subsidiaries or the MNC headquarters.

Motivation as a challenge

Under an organizational division of labour, no agent inside the MNC is likely to have all the knowledge needed for making an optimal choice, and transmitting all of knowledge to him is often prohibitively costly. As Hayek (1945) famously argued, it is a problem of the utilization of knowledge which is not given to anyone in its totality. In response, for example, an MNC management team may trade off the cost of vertical knowledge transfer with the cost of control loss when delegating decision rights to subsidiaries that include subsidiary opportunism and moral hazard (Jensen and Meckling, 1992).

Many scholars (e.g. Kogut and Zander, 1992, 1993; Moran and Ghoshal, 1996; Madhok, 1996) have been critical of the seemingly cynical assumptions made by transaction and agency cost scholars in the MNC literature with respect to human nature and associated motivational challenges. To these critics, opportunism ('self-interest seeking with guile', Williamson, 1996) and moral hazard (i.e. using asymmetric information to one's advantage and the other party's disadvantage) are descriptively inaccurate, theoretically unnecessary in the explanation of governance choice, and eventually misleading in an explanation of the existence, boundaries and internal organization of the MNC.

For example, Kogut and Zander (1993) argue that the efficiency of tacit technology transfer relative to other firms determines boundary choices because cooperation within an organization leads to a set of capabilities that are easier to transfer within the firm than across organizations. Market failure considerations related to conflicting motivation are not required (p. 629): the 'problem with the argument that firms exist due to market failure is that it is over-determined; the assumption of opportunism is not needed, only the differential in costs in the transmission of knowledge within the firm as opposed to between firms'.

On the contrary, Love (1995: 401) asserts: 'Consider the precise reasons why the transfer of uncodifiable, tacit knowledge may be costly. This is because one or other (or both) of the parties has to make a considerable investment to make the knowledge understandable to the transferee.' In other words, parties to a knowledge flow need to be motivated to make specific investments in understanding each other. Moreover, since exchanging partially uncodified knowledge most likely involves high transaction costs in terms of performance measurement or pricing contributions to knowledge flows (Alchian and Demsetz, 1972), incentive problems may arise.

By implication, motivational challenges in addition to cognitive challenges of reaching common understanding among participants to a knowledge flow are an essential element in knowledge governance that may have to be addressed simultaneously in both internal and external organizational design of the MNC. For example, an MNC exploits firm-specific capabilities through own subsidiaries if this mode of exploitation offers higher pay-offs than the licensing out of products to local firms (e.g. Kümmerle, 1999). In other words, the MNC will trade off the costs of internal knowledge flows, converting knowledge into a licensable product (Demsetz, 1988), both in terms of addressing cognitive and motivational challenges, with the knowledge flow costs of establishing a wholly owned subsidiary abroad.

With regard to the internal organization of the MNC, Foss and Mahnke (2003) suggest that the vertical organization of the MNC arises in response to both conflicting interest and asymmetric knowledge. Under an organizational division of knowledge, management may delegate some rights to subsidiaries, ranging from the trivial (to collect and truthfully report accounting data) to the all-important (the right to make strategic decisions in local markets). Management wishes these delegated rights to be exercised in an optimal manner. However, since the subsidiary cannot be constantly monitored and management lacks local knowledge, management either acquires enough local knowledge, which might be cumbersome, or else, delegate decision rights and link their exercise to performance pay schemes. However, because such schemes trade off incentives and risk, some losses (compared to a full-knowledge situation) are usually unavoidable. As a consequence, vertical internal organization arises as a trade-off between these losses, knowledge transfer costs, and the costs of designing monitoring schemes, incentive contracts, and so on (Gupta and Govindarajan, 1991; Jensen and Meckling, 1992).

Another managerial problem of internal organization concerns horizontal relations between an MNC's subsidiaries (e.g. Mahnke and Venzin, 2003). Such horizontal relations can take different forms, including communities of practice, and cross-functional teams etc. Whatever the form, knowledge exchange between subsidiaries cannot be easily forced or directed by the headquarters, but thrives if participants in horizontal knowledge flows are personally motivated to meet cognitive challenges. Similar, Gupta and Govindarajan (2000) suggest that motivation combines with mutual absorptive capacity to enable knowledge flows.

Managerial interference in horizontal knowledge flows between a MNC's subsidiaries to which the top management team has delegated rights (e.g. to seek inter-subsidiary exchange and cross-fertilization possibilities) faces the 'problem of selective intervention' (Williamson, 1996). This is because it is often hard for management to resist interfering. However, arbitrary intervention, the breaking of promises not to intervene, etc., all of which will often be very tempting for the headquarters, are very destructive for motivation (Baker, Gibbons and Murphy, 1999; Foss, 2003).

The bottom line is that a full assessment of what sets of governance mechanisms respond to particular internal and external knowledge flows turns on a number of costs having to do with both cognitive and motivational challenges that have to be balanced against the benefits of various types of knowledge flows.

Governance mechanisms: hierarchy, communities and incentives

Which governance-mechanisms regulate different kinds of knowledge flows? The answer depends, *inter alia*, on cognitive and motivational challenges, in particular knowledge-flow situations. As Grandori (2001) points out, governance mechanisms might be selected according to their ability to address both problems. For example, *horizontal knowledge flows* left to self-organization in inter-subsidiary communities of practice are likely to fail if there is substantial conflict of interest and cognitive partitioning between subsidiaries. Hierarchical coordination, while providing a solution to conflict of interest via fiat also suffers from the 'selective intervention problem' (Williamson, 1996). Accordingly, to align conflicting interests an internal patenting and knowledge market system might be more appropriate because it solves credit assignment problems between subsidiaries. Also in this way, problems of cognitive partitioning can be overcome by exchanging solutions between subsidiaries rather than the knowledge on which they are based (Demsetz, 1988).

Cognitive partitioning, including conflicting judgements of business prospects, impedes *vertical knowledge flows*, for example, between the MNC's headquarters (HQ) and a subsidiary. This is because knowledge sharing by direction (Conner and Prahalad, 1996) is not feasible when the HQ lacks local subsidiary knowledge and vice versa. In addition, extensive lobbying of top management by subsidiaries to adopt their perspectives leads to inefficiency due to bargaining costs (Milgrom, 1988). If this is so, the headquarters might choose to delegate decision rights to the subsidiary to co-locate decision-making power and decision-relevant knowledge accompanied by financial control instead. But at the same time the use of high-powered incentives, such as financial control, is regarded as complicated if measuring problems obtain that, for example, stem from knowledge-flow interdependencies between subsidiaries (e.g. Ouchi, 1980; Gupta and Govindarajan, 1991). Ouchi (1980) suggests that poorly understood cause–effect relations and uncertainty result in ambiguities of performance evaluation – particularly if tasks are highly interrelated.

Only if performance ambiguity is low does output-based performance pay seem to be effective in aligning conflicting interests. Otherwise variable rewards might be appropriate, if pay and control can relate to specified behaviour or to other forms of standardization (e.g. processes), which can serve as a basis for measuring performance. Unfortunately, to the extent that standardization of subsidiary behaviour or processes is prevented, neither behaviours nor outputs can be determined with precision. In this case, Ouchi (1980) suggests, clan control might be the solution to promote cooperation and mitigate conflict of interest: the basis of control becomes a set of internalized values and norms. Bartlett and Ghoshal (1989) agree when they state that managing differentiated network relations for global learning and efficiency benefits from normative integration.

In sum, then, alternative knowledge flows, which are complicated by cognitive and motivational challenges, require the careful selection of governance mechanisms to create value, whereby the MNC might select knowledge-governance mechanisms according to their ability to overcome the challenges of cognitive partitioning and motivation (cf. Grandori, 2001).

Governing knowledge relations

Collectively, the contributions to this volume develop propositions on three crucial questions: In which kind of situation are different key challenges of governing knowledge flows relevant? What are the relevant contingencies influencing knowledge governance in the MNC? How do sets of governance mechanisms respond to these key challenges given particular contingencies? As indicated in Figure 1.2 the contributions in this book address knowledge flows and associated governance challenges both in the MNC's internal and external relations.

Mats Forsgren (Chapter 2) deals with *the use of network theory in MNC research* as a *fundamental research perspective*. It examines how the conceptualization of the MNC has changed since the 1970s. Today the

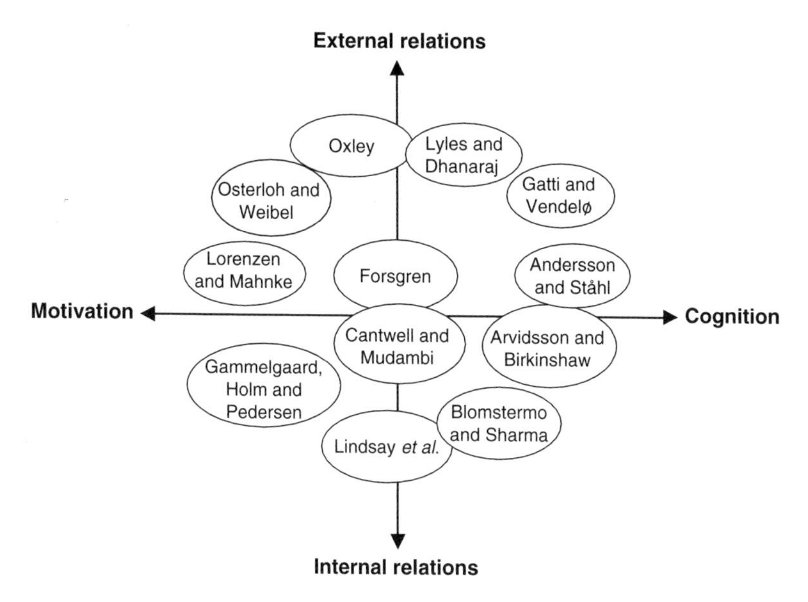

Figure 1.2 Knowledge governance in external and internal relations

conceptualization stems from the MNC controlling a network of global flows of information, capital and people. The network of external and internal relations, rather than the size of the MNC's resources as such, are the focus of value creation. The essay critically reviews the use of the network concepts based on contingency theory, social capital theory, and business network theory. It then proceeds to integrate current research streams and stimulate future empirical research on knowledge governance in the MNC from the perspective of business network theory. *John A. Cantwell and Ram Mudambi* (Chapter 3) characterize and classify various kinds of *competence-creating knowledge flows* to outline how firm's strategic behaviour in managing such flows (both intra-firm and inter-firm) leads to value creation. By integrating perspectives from strategic management (e.g., the knowledge-based view of the firm) and economics (e.g., private good and public good aspects of knowledge), the authors show that the critical element of the analysis is to focus on the MNC's ability to leverage its internal network of subsidiaries in order to integrate knowledge bases that are geographically dispersed. Understanding how to integrate internal and external network relations through knowledge governance appears as crucially important for an MNC's attempts at value creation. Before integration can proceed, however, *external and internal governance challenges* need to be investigated separately.

Governing external knowledge relations

Focusing on external network relations *Margit Osterloh and Antoinette Weibel* (Chapter 4) investigate *social dilemmas in networks*. They explain in great detail the reasons for and possibilities of solving social dilemmas. For example, a social dilemma in an MNC's network arises if interdependent partners face the problem that private and common benefits cannot be optimized jointly, and the optimization of the rather short-term private benefits yields outcomes that leave all partners worse off than would the feasible alternatives. Building on two previously neglected theoretical approaches in MNC research, the public choice and the crowding-out theory, it is shown how the understanding of social dilemmas can contribute to solve motivational knowledge-governance problems. Even if motivational challenges are less of a concern, cognitive challenges remain. *Marjorie Lyles and Charles Dhanaraj* (Chapter 5) comprehensively examine *learning across borders through international alliances from an organizational learning perspective*. Learning has moved from a marginal role in the 1980s to a central role today in managing alliances and joint ventures. Acquiring new knowledge requires distinct learning capabilities as well as an organizational context conducive to dealing with cognitive challenges. Learning even within a singular organization poses formidable challenges, and learning in the context of an interorganizational relationship across national boundaries poses even more complexities. In this chapter, the authors expand the first author's classical contribution (Lyles, 1988) and much research stimulated by it to develop an integrative conceptual framework for organizational learning in international alliances. *Joanne Oxley* (Chapter 6) argues that *learning versus hazard mitigation in alliances may be a false dichotomy*. It has been suggested that attempts by alliance participants to guard against the hazards of opportunism undermine their ability to learn from partner firms. Pitting learning against protection, however, relies on a straw man version of motivational challenges. Critics overlook the central role of 'credible commitments', which enhance learning and protection by discouraging opportunistic behaviour. Important complementarities exist between learning and protection, and this chapter explores possible future directions for research on learning alliances that integrate motivational and cognitive knowledge-governance challenges. *Ulf Andersson and Benjamin Ståhl* (Chapter 7) empirically examine *relationship dynamics and value creation* with a structural model. Despite the existence of multiple theoretical frameworks on why firms engage in business relationships, there is a lack of large-scale empirical research concerning how relationships are developed. The paper finds that knowledge governance benefits from mechanisms enabling task-driven co-action that may be understood as arenas for continuous externalization of the ideas discussed in particular business relationships. The results suggest that such knowledge governance mechanisms are instrumental in fostering knowledge flows and help to transform client–supplier interaction into market-based intellectual assets.

Governing internal knowledge relations

Niklas Arvidsson and Julian Birkinshaw (Chapter 8) suggest that an important prerequisite for the governance of internal knowledge flows concerns the ability to *identify leading-edge market knowledge in the MNC*. Capabilities developed locally are often sticky, hard to identify, and do not flow easily between subsidiary units. It is therefore the combination of external differentiation and sticky internal knowledge flows that leads to the emergence and perseverance of the variation in marketing capabilities. This empirical paper shows how dispersed marketing capabilities can be identified and measured to facilitate internal knowledge flows. *Valerie Lindsay, Doren Chadee, Jan Mattsson and Robert Johnston* (Chapter 9) are concerned with *knowledge flows in international service firms*. They point out that the individual's role in knowledge generation and decision-making in the internationalization process of firms has been largely overlooked. The issue of individual involvement in knowledge flows has particular relevance for service firms, because they tend to rely heavily on embedded knowledge and people as key resources. Often negative personality traits emerge, such as ego defence, jealousy, territorial protection, etc., to pose a formidable motivational challenge that deserves attention when designing knowledge governance mechanisms. *Jens Gammelgaard, Ulf Holm and Torben Pedersen* (Chapter 10) empirically investigate *dilemmas of subsidiary transfer of knowledge*. Because knowledge transfer is demanding in terms of the time and resources necessary, the MNC subsidiary has to decide whether knowledge should be acquired externally or internally and to what extent should the limited resources be spent on external knowledge acquisition versus internal knowledge transfer? The results presented in this chapter confirm that dilemmas are involved in the internal transfer of knowledge from subsidiaries to other MNC units. However, the results also indicate that MNC headquarters can alter the incentive structure to hinder the subsidiary in becoming too locked onto the local context in their knowledge development by deploying several governance mechanisms, including recognition of subsidiary knowledge and granting subsidiary autonomy. While governing external and internal knowledge relations is challenging enough, when external and internal relation change during an MNC's expansion strategy special attention to context contingencies is required.

Knowledge governance and business development

Mark Lorenzen and Volker Mahnke (Chapter 11) develop a proposition on *governing MNC entry in regional knowledge clusters.* MNCs engaged in high-tech business development increasingly seek to tap into agglomeration economies (e.g. specialized labour, vertical and horizontal knowledge spillover) present in geographically dispersed centres of excellence. Entry success in terms of protection against expropriation hazards and internalized agglomeration economies depends on how particular entry modes (greenfield investment, M&A, and strategic alliances) address social and cognitive entry barriers. Cluster-configuration and MNC-specific contingencies are introduced to show when knowledge-seeking MNC entry in knowledge clusters through acquisition, strategic alliances, or greenfield investment is likely to be successful.

Anders Blomstermo and Deo Sharma (Chapter 12) empirically examine *learning and networking in foreign market entry of service firms.* The purpose of the chapter is to investigate the role of experiential learning and knowledge management in service firms expanding in foreign markets. Findings indicate that service firms need to develop a cognitive framework to identify, collect, interpret and transform experience into knowledge. However, there are significant differences within service industries as to what regards knowledge management. Differences in knowledge governance of service firms appear to be contingent on the nature of the service (i.e. hard *vs* soft service), and the internationalization strategy pursued (client-following *vs* market-seeking strategy). Finally, *Anna Gatti and Morten Vendelø* (Chapter 13) comparatively assess *networking by venture capitalists in the US and Europe.* Adopting a cognitive learning perspective, the chapter focuses on both the formation and maintenance of alternative relational ties in networks of meaning creation. A key issue here is not only to understand the structure of the network but also to understand the process of networking. Networking processes and networks as source of information (or competence, in the case of plug-in behaviour) require a sound understanding on how network actors create meaning for decision-making on technological investments.

Future research on knowledge governance in the MNC

Taken together, the chapters in this volume show that advancing research on governing knowledge flows in the MNC proceeds on several levels of analysis and through a variety of methods. To stimulate further empirical research, the chapters identify key contingencies and trade-offs in the governance of knowledge in the MNC. Contingencies influencing how trade-offs in knowledge governance are resolved through particular governance mechanisms include knowledge-types, knowledge-strategies pursued, regional strategy followed, and the costs of knowledge loss, motivation and communication while applying alternative governance mechanisms. Importantly, however, future integrative research on governing internal and external knowledge flows must proceed by addressing the following empirical issues. Given particular constellations of cognitive and motivational challenges what are the key empirical differences, if any, in the MNC's internal and external knowledge flows? When is it easier to govern knowledge flows within rather than across the boundaries of the MNC? When should headquarters give more attention to addressing cognitive than motivational challenges? These and other questions concerned with the performance impact of a variety of knowledge-governance configurations will be a substantial and enduring concern of many scholars and managers of the modern MNC.

References

Alchian, A. and Demsetz, H. (1972) Production, information costs, and economic organization. In A. Alchian (1977) *Economic Forces at Work*. Indianapolis: Liberty Press.

Argote, L. (1999) *Organizational Learning: Creating, Retaining, and Transferring Knowledge*. Boston: Kluwer Academic.

Baker, G., Gibbons, R. and Murphy, K. J. (1999) Informal authority in organizations. *Journal of Law, Economics and Organization*, 15: 56–73.

Bartlett, C. and Ghoshal, S. (1989) *Managing across Borders: The Transnational Solution*. Boston: Harvard Business School Press.

Boisot, M. H. (1998) *Knowledge Assets: Securing Competitive Advantage in the Information Economy*. Oxford: Oxford University Press.

Buckley, P. J. and Carter, M. J. (1999) Managing cross-border complementary knowledge: conceptual developments in the business process approach to knowledge management in multinational firms. *International Studies of Management and Organization*, 29(1): 80–104.

Cohen, W. M. and Levinthal, D. A. (1990) Absorptive capacity: a new perspective on learning and innovation. *Administrative Science Quarterly*, 35(1): 128–52.

Conner, K. R. and Prahalad, C. K. (1996) A resource-based theory of the firm: Knowledge versus opportunism. *Organization Science*, 7(5): 477–501.

Demsetz, H. (1988) The theory of the firm revisited. *Journal of Law, Economics and Organization*, 4: 141–61.

Foss, N. (2003) Selective intervention and internal hybrids: interpreting and learning from the rise and decline of the oticon spaghetti organization. *Organization Science*, 14(3): 331–50.

Foss, N. and Mahnke, V. (2003) Knowledge management: what can organizational economics contribute? In M. Lyles and M. Esterby (2003) *Handbook of Organizational Learning and Knowledge Management*. Oxford: Blackwell.

Grandori, A. (2001) Neither hierarchy nor identity: knowledge governance mechanisms and the theory of the firm. *Journal of Management and Governance*, 5: 181–399.

Grant, R. M. (1996) Toward a knowledge-based theory of the firm. *Strategic Management Journal*, 17:109–22.

Gupta, A. K. and Govindarajan, V. (1991) Knowledge flows and the structure of control within multinational firms. *Academy of Management Review*, 16(4): 768–92.

Gupta, A. K. and Govindarajan, V. (2000) Knowledge flows within multinational corporations. *Strategic Management Journal*, 21(4): 473–96.

Hamel, G. (1991) Competition for competence and inter-partner learning within international strategic alliances, *Strategic Management Journal*, 12 (special issue): 83–103.

Hayek, F. A. (1945) The use of knowledge in society. In *idem* (1948) *Individualism and Economic Order*. Chicago: University of Chicago Press.

Hedlund, G. (1986) The hypermodern MNC – a heterarchy? *Human Resource Management*, 25(1): 9–35.

Jensen, M. C. and Meckling, W. H. (1992) Specific and general knowledge and organizational structure. In L. Werin and H. Wijkander (eds), *Contract Economics*. Oxford: Blackwell.

Kogut, B. and Zander, U. (1992) Knowledge of the firm, integration capabilities, and the replication of technology. *Organization Science*, 3: 383–97.

Kogut, B. and Zander, U. (1993) Knowledge of the firm and the evolutionary theory of the multinational corporation. *Journal of International Business Studies*, 24: 625–45.

Kümmerle, W. (1999) The drivers of foreign direct investment into research and development: an empirical investigation. *Journal of International Business Studies*, 30(1): 1–24.

Lane, P., Salk, J. E. and Lyles, M. (2001) Absorptive capacity, learning and performance in international joint ventures. *Strategic Management Journal*, 22(12): 1139–61.

Love, J. (1995) Knowledge, market failure and the multinational enterprise: a theoretical note. *Journal of International Business Studies*, 26(2): 399–407.

Lyles, M. A. (1988) Learning among joint venture sophisticated firms. *Management International Review*, 28(special issue): 85–98.

Lyles, M. A. and Salk, J. E. (1996) Knowledge acquisition from foreign parents in international joint ventures: an empirical examination in the Hungarian context. *Journal of International Business Studies*, 29(2): 154–74.

Madhok, A. (1996) The organization of economic activity: transaction costs, firm capabilities, and the nature of governance. *Organization Science*, 7(5): 577–90.
Mahnke, V. and Venzin, M. (2003) Governance of knowledge teams in the MNC. *Management International Review*, 3(special issue): 47–68.
Milgrom, P. (1988) Employment contracts, influence activities and efficient organization design. *Journal of Political Economy*, 96: 42–60.
Moran, P. and Ghoshal, S. (1996) Theories of economic organization: the case for realism and balance. *Academy of Management Review*, 21(1): 58–72.
Nonaka, I. and Takeuchi, H. (1995) *The Knowledge-creating Company*. Oxford: Oxford University Press.
Ouchi, W. (1980) Markets, bureaucracies and clans. *Administrative Science Quarterly*, 25: 129–41.
Szulanski, G. (1996) Exploring internal stickiness: impediments to the transfer of best practice within the firm. *Strategic Management Journal*, 17: 27–43.
Williamson, O. (1996) *The Mechanisms of Governance*. Oxford: Oxford University Press.
Zahra, S. A. and George, G. (2002) Absorptive capacity: a review, reconceptualization, and extension. *Academy of Management Review*, 27(2): 185–203.

2
The Use of Network Theory in MNC Research

Mats Forsgren

The network concept and research on MNCs

In contemporary research on MNC the network concept has been used quite frequently (see e.g. Ghoshal, 1986; Johanson and Mattsson, 1988; Bartlett and Ghoshal, 1989; Forsgren, 1989; Ghoshal and Bartlett, 1990; Forsgren and Johanson, 1992; Ghoshal *et al.*, 1994; Malnight, 1996; Ghoshal and Nohria, 1997; Tsai and Ghoshal, 1998; Zander, 1999; O'Donnell, 2000; Andersson *et al.*, 2001, 2002; Kutschker and Shurig, 2002).[1] A closer look at this research reveals that there are some basic differences in how the network concept is used, which partially demonstrates a conflict between two perspectives based on different theory traditions. The first tradition is based on the *contingency theory approach* and focuses mainly on the internal structure of the MNC. The network concept in this research is mainly used as a new form of organization facilitating the exchange of information between geographically and organizationally dispersed units. The second approach applies the theory of 'markets-as-business-networks' in the analysis of the MNC. This approach puts individual business relationships, in which MNC units are involved, at the centre of analysis. In the following discussion we call this approach the *business network theory*.

Both approaches recognize that the difference between subsidiaries, in terms of capabilities, is an important characteristic that every MNC theory must deal with. However, in developing such a theory, the two perspectives use completely different types of networks. In addition, the first approach is basically normative and – in the contingency theory tradition – deals with how the management should structure its organization. The second approach is much more descriptive and focuses on how the inherently loose structure, due to differences in local business networks, affects internal processes within the MNC.

In the present chapter the different approaches will be outlined and analysed in more depth.

Contingency theory and MNC networks

Research about the environment and the organization of MNCs has been largely dominated by the contingency theory approach ever since Stopford and Wells (1972) published their work on the causal link between the environmentally oriented strategies and the formal structure of the MNC. In their work the geographical spread and product diversification were related to structural archetypes such as the mother–daughter structure, the international division structure, the global structure and the matrix structure. The subsequent research on the organizational design of the multinational firm adopted the same basic logic of a causal relationship between the environment, the strategy and the organization (Hedlund and Rolander, 1990). For instance, Egelhoff (1988) based his contingency model on the information-processing view advocated by Lawrence and Lorsch (1967) and Galbraith (1973). In the centre of his analysis is the quality of fit between the information-processing requirement facing the multinational firm, due to the chosen strategy/environment, and the information-processing capacities of the organization. Although this approach emphasizes the fit itself rather than the forces that create the fit (Egelhoff, 1988: 28), similar to the model of Stopford and Wells (1972), it is first of all a functionalistic, top management-driven perspective (Doz and Prahalad, 1993). The top management is assumed to evaluate the environment of the MNC, design the strategy and choose the appropriate organizational structure and control systems.

Research in the same tradition, in which the *network concept* has been used more explicitly, includes work by Bartlett and Ghoshal (1989), Ghoshal and Nohria (1989), Gupta and Govindarajan (1991, 1994), Ghoshal *et al.* (1994) and Ghoshal and Nohria (1997). The latter researchers conceptualize the MNC as a *differentiated network* and explicitly state that their overarching theoretical perspective is contingency theory. However, in contrast to Egelhoff and others, they emphasize that an MNC consists of diverse subsidiaries, operating in unique national environments, which cannot be adequately addressed by a uniform organization-wide structure. At the crux of the analysis is the premise that the structure of the MNC can be understood as a differentiated network, where the network consists of linkages between the headquarters and the subsidiaries, of linkages between subsidiaries, and the subsidiaries' 'local' linkages.

In comparison with other work in the contingency theory tradition, Ghoshal and Nohria (1997) argue that their focus is the required difference *within* the MNC rather than *between* MNCs. The logic of their model is *internal differentiation*. Subsidiaries differ from each other in terms of environmental complexity and access to local resources. This requires an adaptation of control mechanisms used by the corporate headquarters to the needs of each and every subsidiary. A differentiated fit of the management system, rather than the overall fit, is the bottom line.

In addition to usual mechanisms applied by earlier researchers within the contingency tradition, like formalization and centralization, Ghoshal and Nohria (1997) introduce normative integration as a distinctive control mechanism. In contrast to the usual approach, where diversity at the subsidiary level must be mirrored by the mechanisms used, the basic idea here is to use normative integration as a means that *reduces* the impact of diversity. The headquarters, through creating common norms in the MNC, addresses the problem of subsidiaries differing from each other. These norms are supposed to legitimize actions at the subsidiary level in line with the interest of the overall MNC.

Although the conceptualization of the MNC in Ghoshal and Nohria's (1997) analysis differs from earlier work in the contingency theory tradition, mainly through the introduction of the MNC as an internal network, surprisingly little differs from earlier contributions by Egelhoff and others. For instance, the top management-driven perspective is as explicit as ever. It is still the headquarters that is supposed to create the necessary fit between environment, strategy and structure. The top management's ability to implement organizational changes based on the evaluation of environment and necessary strategies is hardly questioned. One can argue that this ability becomes even more crucial in Ghoshal and Nohria's (1997) model, since good knowledge at the top management level is required of *several* environments and *several* local resources. Their model's demand on the top management's information-processing capacity has increased substantially since the perspective was introduced by Lawrence and Lorsch (1967) and Galbraith (1973), as has the difficulty of implementing organizational changes into a highly differentiated MNC.

Another striking feature is that the environment of the MNC, which plays such an important role as an explanatory variable in the model, is defined by the same broad categories as used by earlier researchers, although the multiplicity of environments is stressed. For instance, when Ghoshal and Nohria (1997) relate subsidiary environment to the design variables centralization and formalization, the complexity of the subsidiary environment is summarized into degree of local competition and technological dynamism in different countries, both assessed by respondents at the top management level (on a scale from 1 to 5). A similar approach is used for measuring the resources at the subsidiary level. Compared to earlier research in the contingency theory tradition, the main difference is that the measurement is made at the country (subsidiary) level. However, the concepts used to define the *characteristics* of the MNC's environment are as unspecified as in earlier research. The approach still begs the question of how much is contained in general statements about competition and dynamism in national markets, when neither the subsidiary's products or counterparts in the market are defined. The environment is rather 'faceless'.

The application in MNC research

The network in Ghoshal and Nohria's approach is first of all an intra-organizational *interpersonal network* (Ghoshal and Nohria, 1997: 151). This network is assumed to serve as glue that hold the vast geographically dispersed MNC together. Through this network the different units can coordinate their activities, because the interpersonal ties makes information exchange possible between interconnected units. Or as the authors state: 'In the absence of such informal ties, coordination would break down because organizational members would have to rely solely on formal structural mechanisms, which would quickly get overloaded' (Ghoshal and Nohria, 1997: 151).

Their application of the network concept raises the following issues. First, the quote above indicates that their network is very similar to what has been called the informal part of the organization. The importance of the informal organization for organizational behaviour has been a main issue ever since Roethlisberger and Dickson (1939), founders of the human relations school, published their studies (Scott, 1981). It is a relevant statement that informal, interpersonal linkages are important for exchange of information in an MNC, because of the overload problem. The question is, though, whether using a new concept for an old phenomenon contributes to our understanding of what is going on in an MNC.[2]

Second, Ghoshal and Nohria focus entirely on internal linkages in their model, even though they point out that the model should be extended to include linkages to organizations in the MNC's environment (Ghoshal and Nohria, 1997: 194). This apparent imbalance in focus reflects the tendency to look upon the network as a *management instrument*, amenable to manipulation or influence by headquarters. If the external part of the network is included this perspective becomes even more problematic. It also raises the issue of the inherent contradiction of using an informal dimension of the organization as a management instrument.

Third, the network concept used by Ghoshal and Nohria (1997) does not tell which sub-units are connected and why. It deals totally with the personal aspects of a network, not with the operational or business part of it. The network enables the 'various interconnected parts of the multinational enterprise to coordinate their activities with one another' (Ghoshal and Nohria, 1997: 151). But why they are connected in the first place is not explained by the approach. It is, of course, reasonable to assume that personal linkages reflect business or operational linkages. However, if they are assumed to coincide, the statement that a personal network connects interconnected parts obviously runs the risk of being a tautology.

Fourth, the approach implies that the network concept first of all deals with communication. However, it is not clear if network *is* communication or if it is a *vehicle* for communication. On one hand, the approach holds the view that information flows between sub-units are facilitated through existing personal networks. On the other hand, empirical research in this tradition often seems to equate networking with communication. A good example of the latter is Ghoshal *et al.* (1994), an empirical analysis of the relationship between networks and inter-subsidiary communication in Matsushita and Philips. In this study networks are measured as the time spent in inter-unit meetings, while communication is measured by contact frequency between the same units. A positive relationship between these two variables seems self-evident and, consequently does not hold much information.

Fifth, the approach raises the question what is differentiated in the *differentiated network MNC*. Although the concept seems to imply that networks are differentiated this is actually not the case. The concept *differentiated* refers to the dissimilarities between subsidiaries, in terms of general concepts such as access to local resource, environmental complexity, etc. The network concept refers to intra-organizational, interpersonal linkages, which do not differ, at least not explicitly, between subsidiaries. Paradoxically, the network concept used in Ghoshal and Nohria's approach functions as a mechanism that *decreases* the degree of differentiation in the MNC rather than something that *reflects* differentiation. The notion that subsidiaries' own (external and internal) networks can be important bases for differentiation is totally absent in the model. The expression 'differentiated network' is therefore quite misleading.

To sum up, the use of the network concept in the contingency theory tradition is limited to intra-organizational, interpersonal linkages. It is reasonable to question whether the approach adds much to what has already been said in earlier research about the importance of the informal organization. At least one can argue, in the light of contemporary research on social networks (see e.g. Granovetter, 1985; Burt, 1992; Uzzi, 1997; Dyer and Singh, 1998; Gulati, 1998; Rowley *et al.*, 2000), that the richness of the network concept is not very well reflected in the contingency theory tradition.

MNC networks and social capital

An interesting development of the intra-organizational, inter-personal view on networks has appeared recently in MNC research. By relating to the concept of *social capital* (see e.g. Coleman, 1988; Nahapiet and Ghoshal, 1998; Adler and Kwon, 2002) Tsai and Ghoshal (1998) analyse the relationship between structural, relational and cognitive elements of an MNC's intra-organizational network. These elements are expected to affect the patterns of resource exchange between subsidiaries and the level of product innovation in the MNC. In contrast to the *differentiated network* perspective this approach allows for a much richer interpretation of the network concept. First, it recognizes that trust and trustworthiness must be considered as important dimensions in every network analysis. The level of trust varies between linkages and can only be built up through interactions over time. Consequently, the network is not only defined through the existence of linkages but also partly through the *content* of these linkages. Furthermore, Tsai and Ghoshal (1998) also recognize that every linkage in a network is more or less connected to other linkages. This implies that the value of the social capital a subsidiary has access to through its membership in a network depends on the subsidiary's degree of centrality in the network. A central position means more access to resources from other units than a peripheral position. Tsai and Ghoshal also argue that an MNC's internal network contains a cognitive dimension in the sense that it fosters the normative integration between the subsidiaries.

The application in MNC research

The application of the concept of social capital has contributed fruitfully to the use of the network concept on the MNC. However, some of the limitations we pointed out above still remain. First, the network concept is as internal as in the contingency theory tradition. There is no network outside the MNC, or at least it is not accounted for in the analysis.[3] This is a serious limitation, especially when the influence of the structure and content of the network relationships on value creation in the MNC is focused. The external part of a subsidiary's network is as relevant as the internal part, or even more relevant, if we consider that it is the external network that constitutes the main difference between subsidiaries.

Second, similar to the approach of the *differentiated network* the network first of all implies a set of personal linkages rather than operational linkages (Tsai and Ghoshal, 1998: 469). To the extent business ties are included in the analysis they are treated as the dependent variable. Personal ties create business ties and not vice versa. Why this is the case is not explained and maybe reflects a general tendency by scholars in organization theory to focus a firm's formal and informal organization rather than the core of its business (Barley and Kunda, 1998).

This last notion leads us over to the other strand of research about MNCs that takes as its starting point the *business network theory* approach. While the contingency theory approach focuses on the internal network and how this holds the MNC together, this approach focuses more explicitly on the individual subsidiary's own network, with an emphasis on the external rather than the internal part of the network. The basic question is how these different subsidiary networks impinge on different processes in the MNC. In contrast to the former approach, there is no explicit or implicit assumption of an internal network holding the MNC together, or being used as an instrument to reach such a goal. In the next section this approach will be scrutinized in more depth.

Business network theory and MNC networks

During the last two decades a conceptualization of the market has appeared that focuses on specific business relationships between actors as the basic building blocs (see e.g. Håkansson, 1982, 1989; von Hippel, 1998; Turnbull and Valla, 1986; Johanson and Mattsson, 1992; Axelsson and Easton, 1992; Grabher, 1993; Håkansson and Snehota, 1995; Forsgren *et al.*, 1995; Uzzi, 1996; Ford, 1997; Dyer and Singh, 1998). This approach also became a substantial part of research on multinational firms (see e.g. Larsson, 1985; Johanson and Mattsson, 1988; Forsgren, 1989; Ghoshal and Bartlett, 1990; Forsgren and Johanson, 1992; Malnight, 1996; Björkman and Forsgren, 1997; Holm and Pedersen, 2000; Kutschker and Shurig, 2002).

At the heart of this approach lies the assumption that suppliers and customers are engaged in long-lasting relationships, which they consider important. Empirical data about some 1000 business relationships in European markets showed that most firms operate in markets where a limited number of customers account for a considerable proportion of the firms' sales (Håkansson, 1982; Turnbull and Valla, 1986). The managers often characterized their customer distribution by a 80–20 rule, saying that 20 per cent of the customers take 80 per cent of a firm's sales. In a similar manner the main part of the firm's purchases of inputs came from a limited number of suppliers. They are important as they secure the effective sourcing and marketing, and because they form a basis for the firms' competence development. The business relationships are significant, intangible assets of the firm. The average age of the relationships investigated was fifteen years, and a considerable number of them were much older (Håkansson, 1982).

Business relationships are established and developed by investing time and resources in interaction with each other. Such relationship-specific investments may include adaptation of products, processes and routines. Adaptations are made gradually as a consequence of two firms learning about each other's ways of performing activities. The relationship-investment processes are often mutual (Forsgren and Johanson, 1992: 4). The business relationships are of critical importance for the firm's business, but they are difficult to grasp for an outsider, because they comprise a number of different and complex dependencies – technical, logistic, social, cognitive and economic – between the parties (Håkansson and Johanson, 1988).

Ties to third parties, such as customers' customer, suppliers' supplier, competitors, and public agencies often condition business in a particular relationship. Consequently, markets are more or less stable networks of business relationships (Forsgren *et al.*, 1995). A business network, therefore, is a set of connected exchange relationships between actors controlling business activities. Firms make investment in such networks. The competitive situation of the firm is a matter of the kind of network they operate in – there is a great difference between being an insider or an outsider – and its relation to these networks. The network is the framework which gives the firm both possibilities and constraints in its business.

Business networks differ from social or personal networks by being explicitly coupled to business activities. In a business field, for instance a specific product area, a number of interrelated activities are pursued. Each activity is more or less dependent on the performance of a number of other activities, which must precede or are expected to ensue. These dependencies constitute closely linked activity chains. Over time the activities are modified and adapted to each other, thus increasing their joint productivity and their cohesion. Thus, a change in the performance of one activity may lead to adjustments of the whole chain. Any such business field involves a number of different actors – firms or firm units – and resources which these actors use to perform the activities. The actors are embedded in the wider web of business activities performed in the field through the activity chains, which encompass several actors and several types of resources (Forsgren and Johanson, 1992; Håkansson and Snehota, 1995).

Social or personal bonds are important ingredients of the business relationships in the business network theory. However, contrary to a dominating trait of social network theory (see e.g. Coleman, 1988; Burt, 1992; Granovetter, 1992; Nahapiet and Ghoshal, 1998) the starting point of the analysis is not the personal ties but the business activities and the interdependencies between actors these activities create. Business activities and activity chains are antecedents to personal bonds rather than vice versa, and the interdependencies between business activities can exist independently of individuals and personal ties.

The application in MNC research

In contrast to the use of the network concept related to the contingency theory tradition, researchers in the business network theory tradition have been much more inclined to start their analysis explicitly from the sub-unit level. Different studies have recognized that the special feature of an MNC is its set of different subsidiaries with different economic, institutional and legal contexts (Kogut, 1993). This notion raises the issue not only of what this variety in contexts means for the organization and management of the MNC, but also how these contexts can be conceptualized. General terms such as complexity, dynamism, degree of competition, resource richness do not capture the essential aspects of the subsidiaries' environment. The business network theory seems to offer an interesting approach to that problem and has consequently been applied by some scholars within the international business research area.

In line with this perspective the MNC is conceptualized as a set of 'quasi-firms' – subsidiaries – each and one them embedded in a pattern of unique and idiosyncratic business relationships. Consequently, different subsidiaries are exposed to different demands from local business actors and also differently exposed to new knowledge, ideas and market opportunities. This perspective concretizes the subsidiary's local environment by identifying the relationships that are most important for the subsidiary's business. The analysis focuses the breath and depth of these relationships in terms of type and frequency of interaction, mutual adaptation of activities, age of the relationships etc. In contrast to earlier research the subsidiary environment has got a 'face'.

Another feature of this approach is that the business network concept is also applied *inside* the MNC. In the same way as subsidiaries are related to external business actors through activity chains they are also related to sister units for the same reason. The intra-organizational administrative or personal links, which play such an important role in the contingency theory tradition, are complemented by intra-organizational business relationships. It is argued that it is at least as important to identify which sub-units are related through interdependent business activities as it is to analyse the administrative or personal links between them.

A common feature of research applying the business network theory is its focus on the external business network in which each subsidiary is embedded. In a series of studies during the last decade it is argued that the subsidiaries' network is an important antecedent to different processes and structures within the MNC.

One of the first attempts to formulate a network theory of the MNC, in which the subsidiaries' external networks are included explicitly, is made by Ghoshal and Bartlett (1990). Their conceptualization is strongly influenced by institutional theorists who argue that organizations, as well as sub-units within organizations, also compete for political power (DiMaggio and Powell, 1983). Different subsidiaries have different possibilities to influence the distribution of resources within the MNC, partly depending on the characteristics of the networks in which the subsidiaries are embedded. A central theme in their analysis is that the possibility for the corporate headquarters to counteract the network-based power at the subsidiary level is contingent on the structure of that network. The higher the density of the subsidiary's network, in terms of degree of interconnectedness between the different actors, the higher the subsidiary's relative influence in the MNC. The reason for these is the difficulty of the corporate headquarters to understand and control the web of relationships on which the subsidiary bases its power (Ghoshal and Bartlett, 1990: 95). Consequently, they argue that in certain situations investments and distribution of resources within the MNC will reflect the different interests of the subsidiaries rather than the overall corporate strategy.

A similar perspective is suggested by Forsgren (1989). Influenced by the resource-dependence perspective (Pfeffer and Salancik, 1978; Astley and Sachdewa, 1984) it is argued that the resource on which the subsidiary bases its influence within the MNC can be conceptualized as the business network in which the subsidiary is embedded. The subsidiary's operations are influenced by the overall strategy of the headquarters *and* by this network. These two forces do not have to coincide. On the contrary, the tension between them can be considerable. While the corporate headquarters strive for a common strategy, in which the subsidiary has a specific role, the subsidiaries' interests are much more influenced by requirements of their business relationships. This conceptualization leads to the conclusion that the MNC can be seen as a 'market' for different interests rather than a hierarchically controlled system, or similar to what Ghoshal and Bartlett would call a 'federative context' (Ghoshal and Bartlett, 1990: 84).

This 'dual role' perspective related to business network theory has been used in empirical research on power structure within the MNC. For instance, some studies have focused the relationship between the characteristics of a subsidiary's network and its autonomy and influence respectively (Forsgren and Pahlberg, 1992; Andersson and Forsgren, 1996, 2000). These studies indicate not only that the corporate headquarters compete with the subsidiary's external business network for influence over the subsidiary's behaviour, but also that the network can facilitate the subsidiary's influence over the rest of the MNC.

The business network theory has also been used to analyse why certain subsidiaries develop specialized competence, which can be used by other subsidiaries in the MNC. For instance, in a large international project, focusing so called 'centres of excellence' in MNCs, the subsidiary's business network plays an important role in the analysis (Holm and Pedersen, 2000). A common theme in this research is that the subsidiary's set of demanding external business relationships stimulates the subsidiary's ability to develop specialized and unique competence. The extent to which this competence is transferred to other units within the MNC, though, partly depends on whether the subsidiary is embedded in a network of business relationships with sister units. Therefore, a coherent analysis of the role of the subsidiary in the MNC requires that both the external and the internal part of the subsidiary's network be considered. This approach also points to the interesting issue of a possible trade-off between embeddedness in the external network and internal network, respectively. Or expressed otherwise, is there a contradiction in developing specialized competence and sharing that competence with sister units?

Subsidiary performance and transfer of competence to other subunits have been the focus of various studies. By using the concept of relational embeddedness (Uzzi, 1997; Gulati, 1998; Dacin *et al.*, 1999) the variation between subsidiaries in terms of market performance and their importance for other MNC units' competence development has been studied (Andersson *et al.*, 2000, 2001, 2002). Applied to the MNC context, relational embeddedness refers to the extent to which customers, suppliers, competitors etc. can serve as sources of learning. An underlying idea is that actors who are strongly tied to each other are more capable of exchanging information, and therefore can learn more from each other (Lane and Lubatkin, 1998; Hansen, 1999). Consequently, a subsidiary does not have an equal capacity to learn from its network. Assimilation of new knowledge is facilitated by the closeness of its existing dyadic relationships, that is by the degree of relational embeddedness. These studies indicate that a high degree of relational embeddedness is positively related not only to the subsidiary's own market performance but also to its importance for sister units' competence development.

By combining the resource-based view with business network theory the relative merits of external and internal sources of knowledge have also been studied. For instance, a study of foreign-owned subsidiaries in Denmark indicates that there is a strong interaction effect of internal factors and external business network factors on subsidiary importance (Forsgren *et al.*, 1999). In a similar vein, studies indicates that transfer of knowledge from a subsidiary to sister units is positively affected by a combination of internal and external sources of knowledge used by the subsidiary (Foss and Pedersen, 2001).

Some limitations of business network theory

The use of business network theory has opened up fruitful ways of conceptualizing the environment of the subsidiaries of an MNC. Compared to the 'differentiated network approach' in contingency theory it offers a more substantial signification of the concept 'differentiated'. Subsidiaries differ due to the characteristics of their business networks. These characteristics have a profound impact not only on the subsidiaries' own performance but also on the role it can play within the MNC. The approach also recognizes the usefulness of the resource dependence perspective in MNC research. It offers a practical tool for analysing some of the power bases, which constitute subsidiary autonomy and subsidiary influence within the MNC. It has been suggested that relationships inside and outside the organization are crucial for influence (Doz and Prahalad, 1993), and the business network theory is a distinctive way to analyse such relationships.

There are some apparent limitations, though, that have to be addressed in future research applying business network theory in MNC research. First, although the perspective implies identifying 'physical and business' factors rather than 'organizational' factors, much more has to be done on that issue. Using the mutual degree of adaptation of specific production, development and marketing activities as indicators of closeness in relationships, and not only degree of trust and trustworthiness, is a step in the right direction. However, factors like type of product, type of technology, etc. must also be included more explicitly in the analysis, especially if the issue of *why* some subsidiaries' relationships are closer than others is examined. There is also much more room for in-depth studies of why a certain relationship in a subsidiary's business network is crucial for performance and influence within the MNC.

Second, in relation to the issue of creation and transfer of knowledge within the MNC the approach needs greater elaboration. For instance, in most research the quality of the subsidiary network is conceptualized as an antecedent to both creation and transfer of knowledge. Closeness in business relationships is supposed to be conducive for the subsidiary's assimilation of new knowledge from the environment. This assimilation in its turn will improve the possibility of the subsidiary to be a giver of knowledge to other MNC units. There is a relatively simple 'picking-up-giving-away' metaphor behind this view, even if knowledge is supposed to be processed and adapted along the way. However, the approach could employ a much more realistic and fruitful analysis of what is created and what is transferred, by looking at different types of relationships. For instance, by using Richardson's (1972) distinction between similarity and complementarity of relationships between firms it becomes apparent that the business network theory first of all deals with creation in terms of *problem-solving* in the latter type of relationship. Through close interaction with customers or suppliers problems are identified, which trigger the mutual development of new solutions, for instance new products or new production techniques. The more intensive this interactive process is, the more the two parties can learn about each other's capabilities. The crucial point, though, is that the process is contingent on *dissimilarities* between capabilities rather than similarities, which seems to be what 'combinative capability' is all about (Kogut and Zander, 1992). Typically, this view does not deal with knowledge *transfer* as an important concept, because identifying who is the sender and who is the receiver in an interaction process is more or less beside the point. Rather, learning in complementary relationships is about how both parties develop and adapt their technologies through mutual problem-solving.

Consequently, the usual view that a subsidiary can 'tap' the environment of knowledge through a good 'teacher–student' condition (Lane and Lubatkin, 1998), or by a common prior knowledge/capability and therefore a high absorptive capacity (Cohen and Levinthal, 1990) is not in the centre of the business network theory. New knowledge is not first of all 'picked up' from the business network in which the subsidiary is embedded, but rather developed in close interaction within the network.

Third, most analyses in line with the business network theory equate the subsidiary business network with a set of dyadic relationships. Or expressed differently, relational embeddedness rather than structural embeddedness provides the focus (Gulati, 1998). This is a serious limitation, because the subsidiary's *position* in the network, and not only the content of the individual relationships, is an important dimension when the subsidiary's resources is assessed. This is especially evident in the light of the discussion of the pros and cons of open versus closed networks and weak versus strong ties (Coleman, 1988; Burt, 1992; Walker *et al.*, 1997; Gulati *et al.*, 2002). For instance, the issue should be addressed whether a subsidiary benefits more from external embeddedness in an open network, with a high degree of non-redundant information, or if a closed network with a higher possibility of a cooperative behaviour is more beneficial. Relational and structural aspects might also be complementary such that in a closed network weak ties may be more desirable than strong ties, whereas in an open network, strong ties may be more desirable (Gulati *et al.*, 2002).

Inherent in business network theory, though, is the view that business networks have no definite border. Consequently, network centrality in the overall network is difficult to estimate. This issue must be addressed in the future in order to capture the essence of a subsidiary's network resources. Related to this issue is also the important issue of the possibility of a subsidiary being too embedded in a network with negative consequences for the creation of new knowledge (Uzzi, 1997; Håkansson and Snehota, 1998; Gulati *et al.*, 2002).

Fourth, the business network theory stresses the evolutionary character rather than radical changes in the network (Halinen *et al.*, 1999). One may also argue that most studies using this approach have emphasized customer–supplier networks rather than horizontal networks, in terms of competitor relationships. However, dissolution of relationships, for instance change of suppliers, and not only development of existing relationships, is an important aspect of every business network. Studies of MNC subsidiary networks, therefore, must also include radical changes of business relationships, in order to assess the subsidiary's competence level and strength within the MNC.

Conclusion

At the time when Stephen Hymer's analysis of the MNC dominated the scene (Hymer, 1972), the multinational corporation was conceptualized as a large oligopolistic firm, which exploited its monopolistic advantage abroad, mainly based on resources located in the home country. Gradually, though, the conceptualization of the MNC has changed since the 1970s. Today the conceptualization stems from the MNC controlling a network of global flows of information, capital and people. The network, rather than the size of the MNC's resources as such, is central. The ability to create global networks and to utilize geographically specialized resources through transfer of knowledge between the nodes of the network is at the core of the analyses. In recent studies of globalization in the post-modernistic society this ability is supposed to give the MNC a superior position vis-à-vis most other firms or groups (see e.g. Castells, 1996; Hardt and Negri, 2000).

The frequent use of the network concept in MNC research mirrors this change. However, it is obvious that the present use of the concept reveals apparent differences in terms of theoretical perspectives. In this chapter the use of the network concept, based on contingency theory, social capital theory and business network theory, has been examined. The analysis of the differences was based on two general problems that every conceptualization of the MNC has to consider: *the environment of the MNC* and the *MNC as an organization*. The basic findings of this analysis can be summarized as follows.

First, studies based on contingency theory and social capital theory have limited their application of the network concept to analysing what is inside the MNC rather than to what is outside. In studies inspired by business network theory, on the other hand, the analyses of the different subsidiaries' environment play a main role. The use of the network concept is an important means to define the important characteristics of the environment by identifying important business relationships.

Second, in the contingency theory tradition the network concept is first of all used to analyse the informal organization of an MNC. MNCs have networks because there are different kinds of informal, personal connections between managers in different subsidiaries. These connections are crucial, because they decrease the information-processing difficulties at the corporate level and constitute the glue that keeps the MNC together. In both the social capital theory and business network theory the network concept is used to capture both formal and informal links. In especially the latter approach the analysis also covers other types of intra-organizational networks than inter-personal networks.

Third, the assumption in contingency theory of the intra-organizational network as a glue contains the view of the MNC as a hierarchical, homogenous and top-driven organization, in which the network first of all is a managerial device among others. In studies based on social capital theory and (even more) business network theory there is much more room for the MNC being a heterogeneous entity, with many different loci of power. The networks to which the different subsidiaries have access constitute sources of power, which they can use in the bargaining process within the MNC.

Fourth, a closer look at the *content* of the relationships of the network reveals fundamental differences between the approaches. Most studies in the contingency theory tradition look upon the network as a system of (informal) communication channels. The analysis focuses on the level of communication, in terms of frequency of contacts, rather than on what is communicated or the reason for the communication. In social capital theory the network concept is much more developed. It is used to analyse the structure of the network, its cognitive dimension as well as the characteristics of individual relationships in terms of trust and trustworthiness. At the centre of the analysis is the issue of how these characteristics influence creation and exchange of knowledge in the MNC.

Fifth, business network theory emphasizes the business activities behind the network. Relationships exist because the activities carried out by business actors are interdependent. These interdependencies are first of all reflected in the degree of adaptation of resources on both sides and the importance they attach to each other's business. Personal bonds, information exchange and trust do exist in these relationships. However, in contrast to both the contingency theory approach and the social capital theory, these dimensions do not *constitute* the network. Instead, the network is manifested by the flows of goods and knowledge between business actors. In line with the business network perspective, the closeness of the relationships, in terms of mutual adaptation, is the basic character of a business network. Also in line with this perspective, a subsidiary's business network comprises both external counterparts and sister units and both parts of the network must be analysed with the same kind of tools.

Notes

1. Due to space constraints this chapter does not cover application of the network theories to other issues in international business research, for instance the internationalization process of firms (see, e.g., Eriksson *et al.*, 1997).
2. Ghoshal and Nohria's approach also raises the question why all personal linkages have to be informal.
3. The concept of social capital, though, has been used to analyse the importance of inter-organizational relationships in other contexts (see e.g., Walker *et al.*, 1997).

References

Adler, P. S. and Kwon, S.-W. (2002) Social capital: prospects for a new concept. *Academy Management Review*, 27: 17–40.

Andersson, U. and Forsgren, M. (1996) Subsidiary embeddedness and control in the multinational corporation. *International Business Review*, 5(5): 487–508.

Andersson, U. and Forsgren, M. (2000) In search of centers of excellence: network embeddedness and subsidiary roles in multinational corporations. *Management International Review*, 40(4): 329–50.

Andersson, U., Forsgren, M. and Holm, U. (2001) Subsidiary embeddedness and competence development in MNCs: a multi-level analysis. *Organization Studies*, 22(6): 1013–34.

Andersson, U., Forsgren, M. and Holm, U. (2002) The strategic impact of external networks: subsidiary performance and competence development in the multinational corporation. *Strategic Management Journal*, 23: 979–96.

Andersson, U., Forsgren, M. and Pedersen, T. (2000) Subsidiary performance in multinational corporations: the importance of technology embeddedness. *International Business Review*, 10: 3–23.

Astley, W. G. and Sachdewa, P. S. (1984) Structural sources of intra-organizational power: a theoretical synthesis. *Academy of Management Review*, 9(1): 104–13.

Axelsson, B. and Easton, G. (1992) *Industrial Networks: A New Reality*. London: Routledge.

Barley, S. R. and Kunda, G. (1998) Bringing work back in. Unpublished paper, Department of Industrial Engineering and Engineering Management, Stanford University.

Bartlett, C. A. and Ghoshal, S. (1989) *Managing across Borders: The Transnational Solution*. Boston: Harvard Business School Press.

Björkman, I. and Forsgren, M. (1997) *The Nature of the International Firm*. Copenhagen: CBS Press.

Burt, R. (1992) *Structural Holes: The Social Structure of Competition*. Cambridge, MA: Harvard University Press.

Castells, E. (1996) *The Information Age*. Oxford: Blackwell Publishers.

Cohen, W. and Levinthal, D. A. (1990) Absorptive capacity: a new perspective on learning and innovation. *Administrative Science Quarterly*, 35(1): 128–52.

Coleman, J. S. (1988) Social capital in the creation of human capital. *American Journal of Sociology*, 94: 95–120.

Dacin, T. M., Ventresca, M. J. and Beal, B. D. (1999) The embeddedness of organizations: dialogue and direction. *Journal of Management*, 25(3): 317–56.

DiMaggio, P. J. and Powell, W. W. (1983) The iron cage revisited: institutional isomorphism and collective rationality in organizational fields. *American Sociological Review*, 48: 147–60.

Doz, Y. and Prahalad, C. K. (1993) Managing DMNCs: a search for a new paradigm. In S. Ghoshal and D. E. Westney (eds), *Organization Theory and the Multinational Corporation*. New York: St Martin's Press.

Dyer, J. and Singh, H. (1998) The relational view: cooperative strategy and sources of interorganizational competitive advantage. *Academy of Management Review*, 23(4): 660–79.

Egelhoff, W. G. (1988) *Organizing the Multinational Enterprise: An Information-processing Perspective*. Cambridge, MA: Ballinger.

Eriksson, K., Majkgård, A., Johanson, J. and Sharma, D. (1997) Experiential knowledge and cost in the internationalization process. *Journal of International Business Studies*, 28(2): 337–60.

Ford, D. (1997) *Understanding Business Markets*. London: Dryden Press.

Forsgren, M. (1989) *Managing the Internationalization Process: The Swedish Case*. London: Routledge.

Forsgren, M., Hägg, I., Håkansson, H., Johanson, J. and Mattsson, L.-G. (1995) *Firms in Networks: A New Perspective on Competitive Power*. Uppsala: Acta Universitatis Uppsaliensis.

Forsgren, M. and Johanson, J. (1992) *Managing Networks in International Business*. Philadelphia: Gordon & Breach.

Forsgren, M. and Pahlberg, C. (1992) Subsidiary influence and autonomy in international firms. *Scandinavian International Business Review*, 1(3): 41–51.

Forsgren, M., Pedersen, T. and Foss, N. J. (1999) Accounting for the strength of MNC subsidiaries: the case of foreign-owned firms in Denmark. *International Business Review*, 8: 181–96.

Foss, N. J. and Pedersen, T. (2001) The MNC as a knowledge source: the roles of knowledge sources and organizational instruments for knowledge creation and transfer. Conference Paper, LINK. Copenhagen: Copenhagen Business School.

Galbraith, J. (1973) *Designing Complex Organizations*. Reading, MA: Addison Wesley.

Ghoshal, S. (1986) The innovative multinational: a differentiated network of organizational roles and management processes. Doctoral dissertation, Harvard Business School.

Ghoshal, S. and Bartlett, C. A. (1990) The multinational corporation as an interorganizational network. *Academy of Management Review*, 15(4): 603–25.

Ghoshal, S., Korine, T. and Szulanski, G. (1994) Interunit communication in multinational corporations. *Management Science*, 40(1): 96–110.

Ghoshal, S. and Nohria, N. (1989) Internal differentiation within multinational corporations. *Strategic Management Journal*, 10: 323–37.

Ghoshal, S. and Nohria, N. (1997) *The Differentiated MNC: Organizing Multinational Corporation for Value Creation*. San Francisco, CA: Jossey-Bass.

Grabher, G. (1993) Rediscovering the social in the economics of interfirm relations. In G. Grabher (ed.), *The Embedded Firm*. London: Routledge.

Granovetter, M. (1985) Economic action and social structure: the problem of embeddedness. *American Journal of Sociology*, 91(3): 481–510.

Granovetter, M. (1992) Problems of explanation in economic sociology. In N. Nohria and R. Eccles (eds), *Networks and Organizations: Structure, Form and Action*. Boston: Harvard Business School Press.

Gulati, R. (1998) Alliances and networks. *Strategic Management Journal*, 19(4): 293–317.
Gulati, R., Dialdin, D. A. and Wang, L. (2002) Organizational networks. In J. A. C. Baum (ed.), *The Blackwell Companion to Organizations*. Oxford: Blackwell.
Gupta, A. and Govindarajan, V. (1991) Knowledge flows and the structure of control within multinational corporations. *Academy of Management Review*, 16(4): 768–92.
Gupta, A. and Govindarajan, V. (1994) Organizing for knowledge flows within MNCs. *International Business Review*, 3(4): 443–57.
Håkansson, H. (ed.) (1982) *Industrial Marketing and Purchasing of Industrial Goods*. Chichester: John Wiley.
Håkansson, H. (1989) *Corporate Technological Behaviour: Cooperations and Networks*. London: Routledge.
Håkansson, H. and Johanson, J. (1988) Formal and informal cooperation strategies in international industrial networks. In F. J. Contractor and P. Lorange (eds), *Cooperative Strategies in International Business*. Boston, MA: Lexington.
Håkansson, H. and Snehota, I. (1995) *Developing Relationships in Business Networks*. London: Routledge.
Halinen, A., Salmi, A. and Havila, V. (1999) From dyadic change to changing business networks: an analytical approach. *Journal of Management Studies*, 36(6): 779–94.
Hansen, M. T. (1999) The search–transfer problem: the role of weak ties in sharing knowledge across organization sub-units. *Administrative Science Quarterly*, 44(1): 82–111.
Hardt, M. and Negri, A. (2000) *Empire*. Cambridge, MA: Harvard University Press.
Hedlund, G. and Rolander, D. (1990) Action in heterarchies: new approaches in managing the MNC. In C. A. Bartlett, Y. Doz and G. Hedlund (eds), *Managing the Global Firm*. London: Routledge.
Hippel, E. von (1998) *Sources of Innovation*. Oxford: Oxford University Press.
Holm, U. and Pedersen, T. (eds) (2000) *The Emergence and Impact of MNC Centers of Excellence: A Subsidiary Perspective*. London: Macmillan.
Hymer, S. (1972) The efficiency (contradictions) of multinational corporation. In G. Paquet (ed.), *The Multinational Corporation and the Nation State*. Houndmills: Macmillan.
Johanson, J. and Mattsson, L.-G. (1988) Internationalization in industrial systems: a network approach. In N. Hood and J.-E. Vahlne (eds), *Strategies in Global Competition*. London: Croom Helm.
Kogut, B. (1993) Learning, or the Importance of Being Inert: country imprinting and international competition. In S. Ghoshal and D. E. Westney (eds), *Organization Theory and the Multinational Corporation*. New York: St Martin's Press.
Kogut, B. and Zander, U. (1992) Knowledge of the firm, combinative capabilities and the replication technology. *Organization Science*, 3(3): 383–97.
Kutschker, M. and Shurig, A. (2002) Embeddedness of subsidiaries in internal and external networks: a prerequisite for technological change. In V. Havila, M. Forsgren and H. Håkansson (eds), *Critical Perspectives on Internationalization*. London: Pergamon.
Lane, P. J. and Lubatkin, M. (1998) Relative absorptive capacity and interorganizational learning. *Strategic Management Journal*, 19: 461–77.

Larsson, A. (1985) *Structure and Change: Power in the Transnational Enterprise*. Uppsala: Acta Universitatis Uppaliensis.
Lawrence, P. R. and Lorsch, J. W. (1967) *Organization and Environment: Managing Differentiation and Integration*. Boston: Harvard Business School Press.
Malnight, T. W. (1996) The transition from decentralized to network-based MNC structures: an evolutionary perspective. *Journal of International Business Studies*, 27(1): 43–65.
Nahapiet, J. and Ghoshal, S. (1998) Social capital, intellectual capital, and organizational advantage. *Academy of Management Review*, 23(2): 242–66.
O'Donnell, S. W. (2000) Managing foreign subsidiaries: agents of headquarters or an interdependent network? *Strategic Management Journal*, 21: 525–48.
Pfeffer, J. and Salancik, G. (1978) *The External Control of Organizations*. New York: Harper & Row.
Richardson, G. B. (1972) The organization of industry. *The Economic Journal*, September: 883–96.
Roethlisberger, F. J. and Dickson, W. J. (1939) *Management and the Worker*. Cambridge, MA: Harvard University Press.
Rowley, T., Behrens, D. and Krackhardt, D. (2000) Redundant governance structures: an analysis of structural and relational embeddedness in the steel and semiconductor industries. *Strategic Management Journal*, 21: 369–86.
Scott, W. R. (1981) *Organizations: Rational, Natural and Open Systems*. Englewood Cliffs, NJ: Prentice Hall.
Stopford, J. M. and Wells, L. T. (1972) *Managing the Multinational Enterprise: Organization of the Firm and Ownership of the Subsidiaries*. London: Longman.
Tsai, W. and Ghoshal, S. (1998) Social capital and value creation: the role of intra-firm networks. *Academy of Management Journal*, 41(4): 464–76.
Turnbull, P. W. and Valla, J.-P. (eds) (1986) *Strategies for International Industrial Marketing*. London: Croom Helm.
Uzzi, B. (1996) The sources and consequences of embeddedness for the economic performance of organizations: the network effect. *American Sociological Review*, 61(August): 674–98.
Uzzi, B. (1997) Social structure and competition in inter-firm networks: the paradox of embeddedness. *Administrative Science Quarterly*, 42: 35–67.
Walker, G., Kogut, B. and Shan, W. (1997) Social capital, structural holes and the formation of an industry network. *Organization Science*, 8(2): 109–25.
Zander, I. (1999) How do you mean 'global'? An empirical investigation of innovation networks in the multinational corporation. *Research Policy*, 28: 195–213.

3

Multinational Enterprises and Competence-Creating Knowledge Flows: A Theoretical Analysis

John A. Cantwell and Ram Mudambi[1]

Introduction

Historically, multinational enterprises (MNEs) located R&D in their subsidiaries abroad mainly for the purposes of the adaptation of products developed in their home countries to local tastes or customer needs, and the adaptation of processes to local resource availabilities and production conditions. In this situation subsidiaries were dependent on the competence of their parent companies, and so their role was essentially just competence exploiting, or in the terminology of Kümmerle (1999) their local R&D was 'home-base exploiting'. In recent years instead, linked to the closer integration of subsidiaries into international networks within the MNE, some subsidiary R&D has gained a more creative role, to generate new technology in accordance with the comparative advantage in innovation of the country in which the subsidiary is located (Cantwell, 1995; Papanastassiou and Pearce, 1997; Cantwell and Janne, 1999; Pearce, 1999; Zander, 1999). This transformation has led to a quantitative increase in the level of R&D undertaken in at least those subsidiaries that have acquired this kind of competence-creating mandate, and in these subsidiaries there has been a qualitative upgrading in the types of research project away from the purely applied towards the more fundamental; although the research undertaken is generally of an (increasingly) specialized kind, to take advantage of the particular capability of local personnel and the other local institutions with which the subsidiary is connected.

The shift towards internationally integrated strategies within MNEs is partly grounded on a 'life cycle' effect within what have become mature

MNEs, which have now created a sufficient international spread in their operations that they have the facility to establish an internal network of specialized subsidiaries, which each evolve a specific regional or global contribution to the MNE beyond the concerns of their own most immediate market (Cantwell and Piscitello, 1999, 2000). Thus, subsidiaries that began as local market-oriented (import-substituting) units are gradually transformed into more export-oriented and internationally integrated operations. While some of the subsidiaries within such a network may have essentially just a competence-exploiting or an 'assembly' role, others take on a more technologically creative function and the level and complexity of their R&D rises accordingly (Cantwell, 1987).

In this chapter we attempt to characterize the knowledge flows associated with MNE operations and to analyse the firm's strategic behaviour in managing such flows (both intra-firm and inter-firm flows) to maximize value creation. This characterization develops work in international business (Gupta and Govindarajan, 1991, 2000) by integrating perspectives from strategic management (e.g., the knowledge-based view of the firm, Grant, 1996) and economics (e.g., private good and public good aspects of knowledge). The critical element of the analysis is the focus on the MNE's ability to leverage its internal network of subsidiaries in order to integrate knowledge bases that are geographically dispersed. We characterize competence creation at the home site as well as at subsidiary sites (see Figure 3.1) We also examine the implications for the locations where the MNE operates.

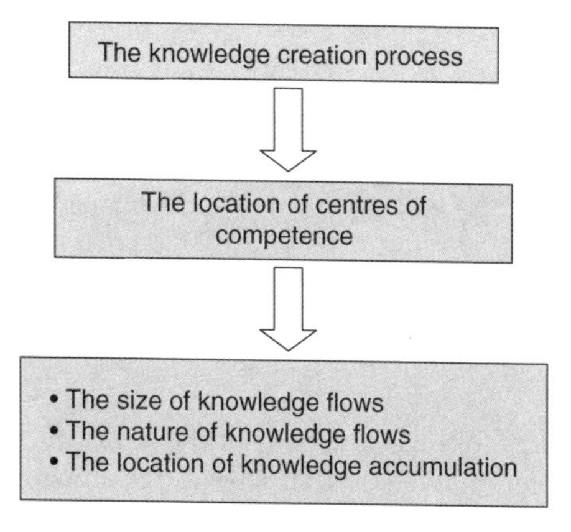

Figure 3.1 Knowledge creation and knowledge flows

The chapter is organized as follows. In Section 2 we trace the history of knowledge processes within MNEs and provide a brief survey of the current literature on geographically dispersed knowledge management. In Section 3 we describe and analyse a model of knowledge flows related to MNE operations. We discuss some implications and offer some concluding remarks in Section 4.

Background

Much of the early literature on knowledge management in the MNE during the post-1945 period viewed it through the lens of the Vernon (1966) product cycle model. Vernon argued that having established a new product or a new production process in the home market, MNEs would subsequently export and/or locate production facilities in foreign locations. This process would inevitably involve some foreign knowledge creation, mainly concerned with adapting the products and the production processes to suit local market conditions. According to this view, MNEs' knowledge management consisted of primary knowledge creation in their home markets and then, as secondary activity, adaptation for use in foreign markets.

A complementary theory is the stage theory of MNE evolution (Johanson and Vahlne, 1977). This has become the textbook view of the MNE and is that of a mature divisionalized company that often grows large in its domestic markets before turning multinational through a sequence of steps beginning with exports. On the basis of this theory, it has been argued that even secondary adaptive knowledge creation in foreign markets occurs at a later stage of foreign operations (Ronstadt, 1977).

Beginning in the late 1970s, even Vernon (1979) observed a changed pattern of MNE knowledge management. Particularly in high technology industries, he suggested that the product cycle had by that time become highly compressed so that many MNEs were engaged in programmes of almost simultaneous knowledge creation in many major markets. This finding has been supported and amplified by a number of other studies, including Cantwell (1995), Dunning (1992) and Howells (1990), though some conflicting evidence has also been presented (Patel and Pavitt, 1991).

In the decades following 1945, home country strengths were exploited in foreign markets so that knowledge activities in these markets were largely driven by local demand. Even prior to 1980, this demand-led view of foreign knowledge management was argued by some to be one-sided and potentially misleading (Mowery and Rosenberg, 1979). More recently, a number of studies point to MNEs adopting a much more sophisticated approach to knowledge management. Considerable evidence has been gathered indicating that the hierarchical model of knowledge management, where strategic decisions were taken at the home country headquarters and suitable implementation decisions were taken in the host country subsidiary, has become encompassed within a more general model.

Forces underlying subsidiary evolution

This change in MNE knowledge management practices has not come about spontaneously, but is the product of powerful external forces. On the demand side, increasing wealth, even in many emerging markets has led to a growth in the demand for more customized products. This has led to MNEs striving for 'mass customization' (Kotha, 1995), where the firm continues to use its home country expertise by exploiting economies of scale and scope, while at the same time incorporating the scope for considerable country-specific differentiation. Customization inevitably involves a large service component that is most effectively delivered at the local or subsidiary level (Pine *et al.*, 1993). This has led to major changes in the role of subsidiaries in the knowledge management process. It has created a need for strategic decision-making at the subsidiary level[2] as subsidiaries are involved more deeply and at a much earlier stage in the MNE's innovation process. In turn, this requires and has led to increased cooperation between headquarters and subsidiaries (Hitt *et al.*, 1998).

On the supply side, many of the information and communication technologies (ICT) have greatly reduced the advantages of mere size. Knowledge, including market knowledge, is increasingly becoming the key source of competitive advantage (Bettis and Hitt, 1995). This has required MNEs to lean ever more heavily on their subsidiary network continually to maintain their knowledge advantage. It has been said that 'in an increasingly networked world economy, it is not the big that will beat the small, but the fast that will beat the slow' (Chambers, 1998).

Finally, the institutions of the world economy are inexorably moving toward increasing the level of competition (Hitt *et al.*, 1998). As trade barriers are rolled back, MNEs find that their former 'profit sanctuaries' are no longer secure and many of their cross-subsidization based advantages have been dissipated. Increased competition and falling rates of return on their traditional activities have forced MNEs to ratchet up their offerings. Thus, forces of demand, supply and institutional change have all pushed MNEs towards becoming more decentralized knowledge management systems.

Intra-MNE knowledge flows

How have these worked in practice? A number of studies have attempted to shed light on this question. Gupta and Govindarajan (2000) report that knowledge flows into and out of subsidiaries depend crucially on the motivation of the subsidiary to acquire knowledge and to share it. This places a great deal of emphasis on firm organization, where the incentive structure of unit managers needs to be carefully designed.

The nature of the knowledge itself has a bearing on how it is managed (Cantwell and Santangelo, 1999). Thus, the norm is for codified knowledge that is relatively easy to transmit to be geographically dispersed, while highly tacit knowledge tends to remain localized. However, in some cases highly tacit knowledge is sourced from foreign subsidiaries and this tends to be driven either by particularly strong and unique local competencies or by particularly strong company-specific networking capabilities. This finding complements and qualifies earlier findings reported by Jaffe *et al.* (1993) that knowledge creation tends to be highly localized.

Further, the position of the MNE in knowledge hierarchy is also likely to have a bearing on its knowledge management practices. Cantwell and Janne (1999) report that MNEs emanating from the leading technological centres in their industries are likely to disperse their knowledge-creating activities, while those from weaker centres of the same industries tend to replicate their technologies in new locations. The subsidiary's location also matters, since the more munificent the location, the greater the potential knowledge creation (Cantwell and Iammarino, 1998).

Cantwell and Piscitello (1999) report that by the 1980s MNEs had begun to utilize their international networks to access locationally specialized branches of innovation across national boundaries. However, they find a trade-off between geographical and sectoral dispersion of knowledge-creating activities. Thus MNEs whose knowledge development spans a wide spread of technological activities tend to have knowledge networks that are less geographically dispersed, and vice versa. This points to the highly complex nature of knowledge-management processes.

Intra-MNE politics

It is important to bear in mind that MNEs are not mechanistically driven towards value maximization. Indeed, the loosening of the traditional hierarchical structure and increased subsidiary role has made them more like political coalitions and less like military formations (Holm and Pedersen, 2000). This implies that a subsidiary's competence development also creates tensions. Indeed, such development may not always strengthen the subsidiary's position within the MNE and could actually hinder it. Forsgren and Pedersen (2000) report that greater subsidiary competence only strengthens its role within the MNE to the extent that other units are able to assimilate and use it. Further, greater effort spent in developing its own competencies (at the expense of engagement in the transfer and use of this competency in other units) actually has a negative effect on the subsidiary's position within the MNE (Forsgren *et al.*, 2000).

Zander (1999) finds that the structure of knowledge networks differs significantly across industries and also amongst firms, concluding that there does not appear to be a single approach to knowledge management. Industry-and firm-specific factors appear to have a significant role in influencing the strategies that are implemented. Further, he also casts doubt on a systematic link between the geographical dispersion of knowledge capabilities and firm performance. He is therefore doubtful whether a strategy of tapping into local competencies is a sure route to developing competitive advantage.

However, a joint reading of Zander (1999) and Forsgren and Pedersen (2000) suggests that implementation and not the strategy may be the problem. In other words, subsidiaries embedded in leading technological centres of competence (Cantwell and Janne, 1999) may be sources of *potential* competitive advantage that remains unrealized due to the internal political structure of the MNE. Indeed, this is very likely to be the case. In a large panel of US firms, Rajan *et al.* (2000) show that when intra-firm diversity becomes high enough, inter-divisional rivalries lead to enormous value destruction through resource transfers from stronger to weaker units. MNEs with geographically dispersed competencies fit the Rajan *et al.* (2000) definition of high-diversity firms extremely well. This lends force to the conclusion of Forsgren and Pedersen (2000) that actually realizing the benefits of the MNE's geographically dispersed knowledge network is a daunting task.

Characterizing knowledge flows related to MNE operations

Our objective is to set up a model to characterize the knowledge flows created within and contiguous to the operations of a modern MNE. As we have seen in the preceding section, two sets of inter-related forces

Table 3.1 Knowledge flow configurations

Case	*Level of knowledge flow*			*Comments*
	Flow 1	*Flow 2*	*Flow 3*	
A	H	H	H	Competitive market structure
B	H	H	L	Subsidiary with high absorptive capacity
C	L	H	H	Competitive market structure
D	L	H	L	Subsidiary with high absorptive capacity
E	L	L	H	Market-seeking FDI
F	L	L	L	Oligopolistic market structure
G	H	L	H	Subsidiary as flagship firm with strategic independence
I	H	L	L	Oligopolistic market structure

have led to a fundamental change in the nature MNE operations. Induced by external drivers and taking advantage of their structure in the face of a changing external environment, MNEs have generally adopted a decentralized approach to their knowledge management processes. Knowledge flows themselves are underpinned by the knowledge creation process and the location of centres of competence within the MNE (see Table 3.1).

We analyse five inter-related aspects of knowledge flows. These are (a) the source–target nature of the flow; (b) the size of the flows relative to a common numeraire; (c) the market structure within which the flows occur; (d) the role of the knowledge in the firm's business process and (e) the applicability of the knowledge.[3] We proceed to discuss each one of these dimensions in detail. We then attempt an integrative analysis of all dimensions. We use this analysis to draw out implications for MNE knowledge strategies, as well as for public policy towards subsidiaries.

The source–target dimension

This dimension is concerned with the nodal and dyadic aspects of knowledge transfer. Thus, each knowledge flow occurs between a source and target along a channel (Gupta and Govindarajan, 2000). We will principally be concerned with three knowledge flows (see Figure 3.2):

Flow 1: Flows from subsidiary to parent. These flows may be called knowledge transfer, and form the basis of the MNE's network leverage. High levels of these flows enable MNE headquarters to exploit

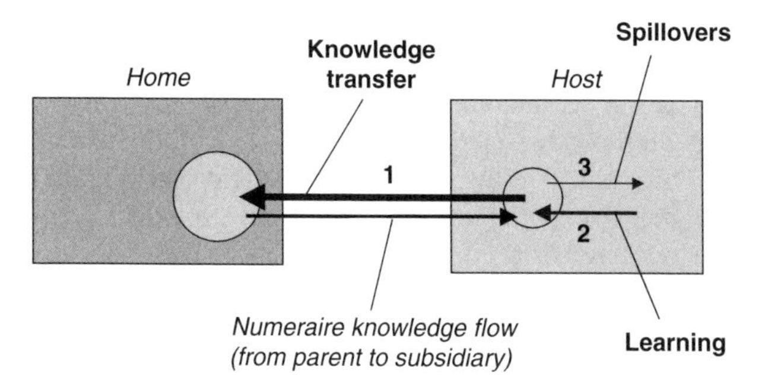

Figure 3.2 Types of knowledge flow

local competencies and act as a knowledge intermediary or knowledge integrator (or as a facilitator of such activities).
Flow 2: Flows from location to subsidiary. These flows consist of the subsidiary's learning, local competence exploitation and local resource utilization. This is the pod or 'listening post' role of the subsidiary where the receiver competence (assessing, filtering and choosing information) and absorptive capacity (adapting inflows to fit firm-specific requirements) become crucial.
Flow 3: Flows from subsidiary to location. These flows are part of what have been termed spillovers. In the literature, spillovers have often been used to refer to flows both into and out of the firm. However, since our analysis is firm-centric rather than location-centric, we define spillovers to include only outflows from the subsidiary. Further, we recognize that spillovers include both intended and unintended elements. Thus, flows to local partners, customers and suppliers may be largely planned. However, flows that occur through employee mobility, local imitation and reverse engineering by competitors may be largely unintentional. The combination of these flows are the focus of most public policy measures, like those implemented in the context of national systems of innovation (NSIs). For example, Fransmann (1997) analyses the factors underlying the success of such policies adopted by MITI in Japan and Mudambi (1998) suggests that the absence of some of these factors may explain the relatively poor results of such policies in Europe.

The fourth knowledge flow from the parent to the subsidiary is the traditional flow, where the subsidiary exploits a home-base knowledge advantage. We do not mean to minimize the importance of this flow, but rather will normalize all flows relative to this flow, i.e., this traditional flow is our numeraire. Any of the three flows above will be described as high or low relative to the knowledge flow from parent to subsidiary. Thus, in the terminology of Gupta and Govindarajan (1991), if Flow 1 is high, then the subsidiary is a net provider of knowledge, whereas if Flow 1 is low then it is a net user of knowledge.

Knowledge flows, whether intentional and unintentional, flow through channels and the nature of these channels affects the quality and quantity of knowledge received by the target. For most channels, knowledge flows are subject to transmission losses in sense of Shannon and Weaver (1998). This is equally true whether the flows are within the MNE or are across firm boundaries. This means that in general the flow of knowledge received by the target will be smaller than the flow transmitted by the source. A number of factors influence the level of transmission losses. For example, transmission losses are greater for tacit knowledge than for codified knowledge (Roberts, 2000).

However, as discussed by Nonaka and Takeuchi (1995), the problem of transmission losses can be solved using rich communication media. These media comprise face-to-face communication, informal interaction and team-based mechanisms (Daft and Lengel, 1986). Using rich communication media can reverse transmission losses, so that the communication process itself results in an enrichment of the original knowledge flow. Thus, the target receives a larger knowledge flow than the originating flow from the source. However, rich communication media are typically high-cost media, so that the firm faces an optimization problem in its choice of knowledge channel, i.e., is the high cost justified by the improved knowledge transfer?

The knowledge channel choice is also influenced by the nature of the knowledge that is being transferred. Szulanski (1996) points out that tacit, context-specific and ambiguous knowledge is likely to be the most difficult to transfer within the firm. This implies that in conventional mechanistic channels, codified knowledge is likely to be subject to low transmission losses in whereas tacit knowledge is likely to be subject to high transmission losses. Thus, rich communication media are likely to be used in transferring such knowledge.

However, when knowledge flows through nodes, in addition to pure transmission issues, motivational considerations (both strategic and environmental) become relevant (Szulanski, 1996). Thus, Flow 1 from the subsidiary to its parent will in general be different from Flow 2 from the location to the subsidiary. This point will be amplified below.

The size of knowledge flows

In the context of knowledge flows, the problem of measurement is acute. How do we measure knowledge? This is particularly problematic when the knowledge itself is composed of both codified and tacit components. Previous studies (e.g. Jaffe *et al.*, 1993; Cantwell and Piscitello, 1999, 2000) have used quantitative patent data to measure codified knowledge flows. Tacit knowledge flows have generally been inferred from questionnaire data (e.g. Håkanson and Nobel, 1993). Neither of these measurement methods relates the size of knowledge flows to value-creation potential. In an interesting study, Hakanson and Nobel (1993) find that the size of knowledge flows from subsidiary to parent (Flow 1) is strongly related to their potential for creating value, i.e., size and value creation are positively correlated. We may conjecture that this positive correlation is likely to hold for most intra-firm flows within the MNE, moderated by some political considerations noted above in Section 2.[4]

For a given source–target pair, the size of knowledge flows will be strongly influenced by the pre-existing knowledge stock at the two ends of the channel. These two stocks determine the 'absorptive capacity' of the target, i.e., the quantity of knowledge that the target can internalize and the speed at which it can do so (Cohen and Levinthal, 1990). The greater the knowledge-stock disadvantage of the target as compared to the source, the lower its absorptive capacity. For example, if the source is much more advanced in its level of technology, most of the knowledge that it can transmit may be beyond the understanding of the target, restricting the size of the possible knowledge flow. Thus, the target's absorptive capacity places an upper bound on the maximal size of the knowledge flow.

All other factors being equal, it is still more difficult to transfer highly tacit knowledge as compared to codified knowledge. Thus, codified knowledge can be transferred in large volumes at relatively low cost. However, large volumes of highly tacit knowledge are difficult to transfer and are only possible at relatively high cost.

The size of knowledge flow is also contingent on the motivation of agents within nodes, i.e., within the subsidiary and within the headquarters of the MNE. Even very favourable knowledge-creation, knowledge-absorption conditions and knowledge characteristics will not lead to large knowledge flows if the agents managing the process are not motivated to ensure that they happen. This motivation is generally a matter of incentive design dictated by the nature of the principal-agent game being played between the MNE headquarters, its subsidiaries and its external knowledge network (Mudambi, 1999). Incorporation of each agent's objectives will generally result in an optimal set of incentives. However, much of the earlier literature has ignored the importance of intrinsic motivation (Frey, 1997) and this can be particularly important in the context of the highly skilled agents involved in knowledge transfer (Osterloh and Frey, 2000).

A closely related point in considering the configuration of knowledge flows is the issue of reciprocity. A large Flow 2 into the subsidiary from the location is often contingent on a large Flow 3 out of the subsidiary into the location. There is considerable evidence both in case study research as well as in more formal game theoretic models that symmetric knowledge flows based on reciprocity are more stable than asymmetric flows.

Given the three knowledge flows under consideration, there are eight possible size configurations. We begin by specifying the configurations and then proceed to analyse each case (see Table 3.1).

Case A: HHH This is the virtuous case with high knowledge transfer, high subsidiary learning and high local spillovers. The subsidiary is likely to enjoy positive interactions with both its parent and with local authorities.

Case B: HHL This is similar to A, but with limited spillovers. This is likely to be the case where the subsidiary as has much to learn but little to contribute. Co-location of a weaker firm into a leading technological centre would fall into this category.

Case C: LHH This case is also similar to A, where the subsidiary is very integrated into the location, but where it finds it difficult to transfer the knowledge it has gained to its parent. The nature of the knowledge (e.g., high levels of tacitness) or the nature of the organization of the firm (e.g., subsidiaries prone to high levels of internecine rivalry as in the analysis of Rajan *et al.* (2000)) can lead to this configuration. The nature of the subsidiary can also be a factor. Too high a level of local integration (Forsgren *et al.*, 2000) where the subsidiary becomes isomorphic to the location (Ghoshal and Westney, 1993) or 'embedded' in it (Andersson and Forsgren, 2000) can increase the distance and reduce trust in the relationship with headquarters, reducing knowledge flow. Obviously any combination of these factors can have the same effect.

Case D: LHL This case is a combination of B and C. Like B, in this case the subsidiary has little to contribute and much to learn. As in B, this is likely to be weaker firm co-locating into a centre of excellence. However, as in C, the knowledge transfer from subsidiary to parent is poor, as discussed above.

Case E: LLH In this case, the subsidiary generates large spillovers, but there is little learning and knowledge transfer to the parent. This case is not the primary focus of this chapter, since it is likely to represent market-seeking FDI, where the MNE's knowledge spillover in the location is compensated by a flow of profit. Adverse selection operates here, since if there are no compensating profits, there is no reason for the firm to place its knowledge-intensive activities in this location.

Case F: LLL This is the opposite of A, where there are virtually no knowledge flows of any kind. This situation is likely to be representative of many firms, although companies may have differentiated knowledge even if they are not at the leading edge of knowledge creation. An example might be a highly replicative industry (like franchise operations) where the knowledge flow consists mostly of a closed business system. Like E, this is likely to be market-seeking FDI.

Case G: HLH This is the opposite of D. The subsidiary generates large spillovers, but it has little to learn from the location. It is nonetheless able to generate large knowledge transfers to its parent. This may be the case of the subsidiary as a flagship firm with substantial strategic independence (Rugman and D'Cruz, 1997). Thus, it is able to synthesize a great deal of knowledge using relatively few local inflows, while at the same time generating large spillovers to its associated peripheral firms. It is simultaneously able to provide substantial knowledge transfers to its parent that can be used to create value elsewhere in the multinational network.

Case I: HLL This case is similar to G, with the difference being that relatively low spillovers occur. This would be the case if the peripheral firms in the area were competitive rather than collaborative and if the firm itself were oligopolistic, so that it would view outflows to competitors negatively and strive to prevent them.

Market structure and knowledge flows

We may assume that all firms regard knowledge inflows (like Flow 2) positively. However, perceptions of knowledge outflows (like Flow 3) may differ, depending on market structure. Knowledge spillovers have both positive and negative effects on the firm. The private effect on the owner firm is a leakage of its valuable intellectual capital, which would be viewed negatively (Grindley and Teece, 1997). The positive effect is the public good aspect of knowledge (d'Aspremont *et al.*, 1998), where outflows contribute to a virtuous cycle by strengthening the knowledge base of the location and making it more attractive to other knowledge-bearing firms. This, in turn, makes for larger knowledge inflows in the future. The firm's view of spillovers (Flow 3) will depend on its assessment of these two effects.

A competitive market structure is characterized by a large number of firms, each with a relatively small market share and profits. Thus such firms have less to lose from knowledge outflows and more to gain from inflows stemming from a strong location, so that the public good aspect of knowledge may predominate and Flow 3 is viewed as an overall positive. On the other hand, an oligopolistic market structure is characterized by a few large firms, each with a large market share and considerable strategic interdependence. Thus, oligopolistic firms realize that knowledge spillovers to industry rivals can be extremely costly in terms of lost competitive advantage, so that the private good aspect of knowledge is the dominant consideration. Consequently, such firms may view Flow 3 as an overall negative.

This analysis implies that market structure should have a strong effect. When the overall effect of knowledge spillovers (Flow 3) is perceived to be negative, leading firms should not locate in clusters.[5] This is because their outflows are very valuable to their competitors, but the inflows from these competitors are less valuable. In this case, adverse selection ensures that clusters of competing firms, become concentrations of mediocrity. Then, in the spirit of Akerlof's (1970) lemons model, clusters including large oligopolistic competitors should fail to form.

Conversely, in relatively competitive market structures, the positive effects of knowledge spillovers may be considered to outweigh the negative effects, so that co-location of leading firms becomes likely. This analysis would explain the empirical finding that the largest firms do not co-locate their knowledge creation activities with those of their competitive rivals (e.g. Cantwell and Santangelo, 1999). It would also explain the co-location of knowledge creation activities in high-quality clusters in competitive industries (Jaffe *et al.*, 1993; Saxenian, 1994).

The above argument is based on the perspective of the incoming MNE. However, an analysis of the perspective of the region is equally important. When the inter-firm dispersion of the innovative activity is high, the regional innovation system is characterized by considerable interaction amongst multiple poles of innovation excellence. This is common in many regions of the UK and Italy. In this case, a local presence is attractive for the MNE as there are plentiful opportunities for the sourcing of local technology. However, when the regional system of innovation is dominated by a single firm (as in many regions of Germany), inter-firm links are largely of the dependent kind, with firms in the local cluster hierarchically linked to the local giant. In this case, technology opportunities for incoming MNEs become more limited.

Knowledge flows and the firm's business process

Thus far we have assumed that all knowledge flows are homogeneous. This is obviously not the case and we must characterize knowledge flows relative to the MNE's business model. In this context we distinguish between core and complementary knowledge flows. Core knowledge relates to the firm's main business and is the basis for its value creation. Complementary knowledge, on the other hand, relates to secondary processes that support the firm's main business. For example, chemical technology represents core knowledge for a chemical firm, while information technology (IT) may relate to its secondary processes. Obviously there is a continuous spectrum between core and complementary knowledge so that knowledge can be said to more or less 'core'.

We can link our analysis of market structure to the nature of the knowledge flow in order to draw out implications for the location of subsidiary knowledge-creation activities. Thus, our analysis of market structure implies that in oligopolistic industries co-location of *core* knowledge-creation activities is unlikely, since outflows to competitors are deemed too costly. However, *complementary* knowledge-creation activities are likely to be co-located with leading firms in these activities since this relationship is likely to be collaborative, i.e., a firm in industry A would co-locate a knowledge-creating subsidiary with a leading firm in industry B, where B's technology is complementary to A. Thus, clusters of oligopolistic firms would be characterized by complementary knowledge flows, while clusters of competitive firms would be characterized by core knowledge flows.

Knowledge flows and knowledge applicability

Another perspective from which knowledge is assessed is the objective distinction with regard to its applicability. Some technologies have wider applicability than others. In this context, we can distinguish between specialized technologies and general-purpose (GP) technologies. Specialized technologies have limited applicability (and value) outside of the firm's specific industry. However, GP technologies have a much wider applicability (Eliasson, 1997; Helpman, 1998). Thus many of the technologies in the chemical and pharmaceutical industries are likely to be highly specialized, whereas many technologies in the IT and machine tool industries are likely to be GP technologies. Most specialized technologies are likely to form core knowledge, since they have limited applicability outside of their 'home' industry. Most GP technologies are likely to function as complementary knowledge in a wide range of industries. Following our analysis, industries whose core knowledge is operationalized in the form of GP technologies are most likely to be characterized by clusters and consequently, larger knowledge flows.

Implications and concluding remarks

In this chapter we model the knowledge flows created by the activities of an MNE. We have attempted to provide a more complete picture of the process underlying such knowledge flows by integrating three perspectives – that of the MNE parent, that of the subsidiary and that of the subsidiary's host site. Early work in international business generally adopted the perspective of the MNE parent, while work in international management tended to adopt the perspective of the subsidiary. Beginning with the pioneering work of Gupta and Govindarajan (1991), attempts have been made to integrate these two perspectives and study knowledge flows within the MNE as flows within a network of nodes. In the regional science literature in-depth studies of spillover flows into the host location were undertaken, but these typically treated the MNE as a single decision-making unit (Young, Hood and Peters, 1994).

While our model underlines the contributions of earlier literature along each of the three perspectives, it suggests that an integrated view of all three perspectives yields some new insights. First, consider the strategy of leveraging an MNE's internal international network by accessing knowledge in the leading technology centres through its subsidiaries there. Earlier literature has suggested numerous issues concerning the development of the subsidiary's absorptive capacity, following the analysis of Cohen and Levinthal (1990). However, while problems in the subsidiary–parent knowledge transmission process have been pointed out (e.g. Forsgren *et al.*, 2000), optimal incentive design to solve this principal–agent game has not been addressed. A joint consideration of the incentives of subsidiary and parent under different organizational designs would go a long way toward addressing this problem. Essentially this requires considering ways of rendering Cases C and D in Table 3.1 behaviourally sub-optimal.

Second, consider the perspective of the host location. Typically a government or quasi-government agency articulates this perspective (e.g. Mudambi, 1999). Previous literature has suggested choice rules for developing the local economic base (Young, Hood and Wilson, 1994). However, an appreciation of the internal dynamics of the MNE would greatly increase the effectiveness of such agencies. This includes both the pre-investment location phase as well as the post-investment knowledge management phase. Examples of these considerations include higher knowledge spillovers and MNE knowledge-intensive investment attraction. Thus, higher local spillovers through local integration may be a short-term benefit if they come at the expense of the subsidiary's engagement with the rest of its intra-firm network. Attempts at attracting fresh MNE investment by leveraging of existing stocks of knowledge-intensive investment in the quest for critical mass must take account of the nature of the industry structure. Attempting to attract competitors from an oligopolistic industry may be ill advised.

Finally, how should the MNE view spillovers? We have suggested that market structure and firm organization both have a bearing on whether the proprietary or public good aspects of knowledge will gain ascendancy within the firm. Thus, when the proprietary aspect predominates, it is likely that adverse selection will ensure that clusters of like firms will fail to form. This explains the finding that in oligopolistic industries co-location of knowledge-creating activities with competing firms rarely occurs. Further, that clusters with such firms consist of the co-location of firms generating complementary knowledge. Conversely, in competitive industries, the public good aspect of knowledge is likely to predominate, leading to the formation of homogeneous clusters that generate knowledge synergies.

Notes

1. *Acknowledgements*. We would like to thank participants at the 2001 LINK Conference in Copenhagen and especially Jetta Frost, Lars Håkanson and Deo Sharma for comments that substantially improved the chapter. The usual disclaimer applies.
2. Customization is understood to include a substantial pre-manufacture design function whereas adaptation is understood to involve post-manufacture activity such as retrofit.
3. The nature of the knowledge itself – codified *vs* tacit – is also an important consideration that will run through the analysis.
4. According to both the resource-based and the knowledge/competence-based views of the firm, value-creation potential comes from the rareness of the resource or competence. In this context we can distinguish between

'defensive' knowledge flows meant to match the industry standard (and ensure survival) and 'offensive' knowledge flows meant to create competitive advantage. Only the latter flows lead to measurable value creation, providing a second moderating influence on the positive correlation between knowledge flows and value creation.

5. We distinguish between clusters and networks as related to the two aspects of spillover knowledge flows. Clusters consist of co-located firms without any planned knowledge sharing so that spillover knowledge flows amongst them are largely unintended. However, networks consist of firms with planned knowledge sharing so that spillover knowledge flows amongst them are intentional.

References

Akerlof, G. A. (1970) The market for 'lemons': quality uncertainty and the market mechanism. *Quarterly Journal of Economics*, 84(3): 488–500.

Andersson, U. and Forsgren, M. (2000) In search of centre of excellence: network embeddedness and subsidiary roles in multinational corporations. *Management International Review*, 40(4): 329–50.

Bettis, R. and Hitt, M. A. (1995) The new competitive landscape. *Strategic Management Journal*, 16: 7–19.

Cantwell, J. A. (1987) The reorganization of European industries after integration: selected evidence on the role of transnational enterprise activities. *Journal of Common Market Studies*, 26: 127–51.

Cantwell, J. A. (1995) The globalisation of technology: what remains of the product cycle model? *Cambridge Journal of Economics*, 19: 155–74.

Cantwell, J. A. and Iammarino, S. (1998) MNCs, technological innovation and regional systems in the EU: some evidence in the Italian case. *International Journal of the Economics of Business*, 5(3): 383–408.

Cantwell, J. A. and Janne, O. E. M. (1999) Technological globalization and innovative centers: the role of corporate technological leadership and locational hierarchy. *Research Policy*, 28(2–3): 119–44.

Cantwell, J. A. and Piscitello, L. (1999) The emergence of corporate international networks for the accumulation of dispersed technological competences. *Management International Review*, 39(special issue 1): 123–47.

Cantwell, J. A. and Piscitello, L. (2000) Accumulating technological competence: its changing impact upon corporate diversification and internationalization. *Industrial and Corporate Change*, 9(1): 21–51.

Cantwell, J. A. and Santangelo, G. D. (1999) The frontier of international technology networks: sourcing abroad the most highly tacit capabilities. *Information Economics and Policy*, 11: 101–23.

Chambers, J. T. (1998) Competing in the Internet economy. *The Wall Street Journal*, Technology Summit, New York.

Cohen, W. M. and Levinthal, D. A. (1990) Absorptive capacity: a new perspective on learning and innovation. *Administrative Science Quarterly*, 35(1): 128–52.

Daft, R. L. and Lengel, R. H. (1986) Organizational and information requirements, media richness and structural design. *Management Science*, 32(5): 554–71.

D'Aspremont, C., Bhattacharya, S. and Gerard-Varet, L.-A. (1998) Knowledge as a public good: efficient sharing and incentives for development effort. *Journal of Mathematical Economics*, 30(4): 389–404.

Dunning, J. (1992) Multinational enterprises and the globalization of innovatory capacity. In O. Grandstrand, L. Hakanson and S. Sjolander (eds) *Technology Management and International Business: Internationalization of R&D and Technology*. Chichester: John Wiley.

Eliasson, G. (1997) General purpose technologies, industrial competence and economic growth. In Bo Carlsson (ed.) *Technological Systems and Industrial Dynamics*. Dordrecht/Boston: Kluwer, pp. 201–53.

Forsgren, M., Johanson, J. and Sharma, D. (2000) Development of MNC centres of excellence. In U. Holm and T. Pedersen (eds) *The Emergence and Impact of MNE Centres of Excellence*. London: Macmillan, pp. 45–67.

Forsgren, M. and Pedersen, T. (2000) Subsidiary influence and corporate learning: centres of excellence in Danish foreign-owned firms. In U. Holm and T. Pedersen (eds) *The Emergence and Impact of MNE Centres of Excellence*. London: Macmillan, pp. 45–67.

Fransmann, M. (1997) Is national technology policy obsolete in a globalized world? The Japanese response. In D. Archibugi and J. Michie (eds) *Technology, Globalization and Economic Performance*. Cambridge: Cambridge University Press, pp. 50–82.

Frey, B. S. (1997) *Not Just for the Money: An Economic Theory of Personal Motivation*. Cheltenham, UK/Brookfield, USA: Edward Elgar.

Ghoshal, S. and Westney, E. (1993) *Organization Theory and the Multinational Corporation*. New York: St Martin's Press.

Grant, R. M. (1996) Toward a knowledge-based theory of the firm. *Strategic Management Journal*, 17: 109–22.

Grindley, P. C. and Teece, D. J. (1997) Managing intellectual capital: licensing and cross-licensing in semiconductors and electronics. *California Management Review*, 39(2): 8–41.

Gupta, A. K. and Govindarajan, V. (1991) Knowledge flows and the structure of control within multinational firms. *Academy of Management Review*, 16(4): 768–92.

Gupta, A. K. and Govindarajan, V. (2000) Knowledge flows within multinational corporations. *Strategic Management Journal*, 21(4): 473–96.

Håkanson, L. and Nobel, R. (1993) Determinants of foreign R&D in Swedish multinationals. *Research Policy*, 22(5–6): 397–411.

Helpman, E. (1998) *General Purpose Technologies and Economic Growth*. Cambridge, MA: MIT Press.

Hitt, M. A., Keats, B. W. and DeMarie, S. M. (1998) Navigating the new competitive landscape: building strategic flexibility and competitive advantage in the 21st century. *Academy of Management Executive*, 12(4): 22–42.

Holm, U. and Pedersen, T. (eds) (2000) *The Emergence and Impact of MNE Centres of Excellence*. London: Macmillan.

Howells, J. (1990) The internationalization of R&D and the development of global research networks. *Regional Studies*, 24: 495–512.

Jaffe, A., Trajtenberg, M. and Henderson, R. (1993) Geographic localization of knowledge spillovers as evidenced by patent citations. *Quarterly Journal of Economics*, 108(3): 577.

Johanson, J. and Vahlne, J. (1977) The internationalization process of the firm: a model of knowledge development and increasing foreign market commitments. *Journal of International Business Studies*, 7: 22–32.

Kotha, S. (1995) Mass customization: implementing the emerging paradigm for competitive advantage. *Strategic Management Journal*, 16: 21–42.

Kümmerle, W. (1999) The drivers of foreign direct investment into research and development: an empirical investigation. *Journal of International Business Studies*, 30(1): 1–24.

Mowery, D. C. and Rosenberg, N. (1979) The influence of market demand upon innovation: a critical review of some recent studies. *Research Policy*, 8: 103–53.

Mudambi, R. (1998) Review of 'Technology, globalization and economic performance'. *Journal of Management Studies*, 35(5): 690–2.

Mudambi, R. (1999) Multinational investment attraction: principal-agent considerations. *International Journal of the Economics of Business*, 6(1): 65–79.

Nonaka, I. and Takeuchi, H. (1995). *The Knowledge-creating Company*. Oxford: Oxford University Press.

Osterloh, M. and Frey, B. S. (2000) Motivation, knowledge transfer and organizational forms. *Organization Science*, 11(5): 538–50.

Papanastassiou, M. and Pearce, R. D. (1997) Technology sourcing and the strategic roles of manufacturing subsidiaries in the UK: local competences and global competitiveness. *Management International Review*, 37: 5–25.

Patel, P. and Pavitt, K. (1991) Large firms in the production of the world's technology: an important case of 'non-globalization'. *Journal of International Business Studies*, 22(1): 1–22.

Pearce, R. D. (1999) Decentralized R&D and strategic competitiveness: globalized approaches to generation and use of technology in MNEs. *Research Policy*, 28(2–3): 157–78.

Pine, B., Victor, B. and Boynton, A. C. (1993) Making mass customization work. *Harvard Business Review*, 71: 108–19.

Rajan, R., Servaes, H. and Zingales, L. (2000) The cost of diversity: the diversification discount and inefficient investment. *Journal of Finance*, 55(1): 35–80.

Roberts, J. (2000) From know-how to show-how? Questioning the role of information and communication technologies in knowledge transfer. *Technology Analysis & Strategic Management*, 12(4): 429–43.

Ronstadt, R. C. (1977) *Research and Development Abroad by US Multinationals*. New York: Praeger.

Rugman, A. and D'Cruz, J. (1997) The theory of the flagship firm. *European Management Journal*, 15(4): 403–12.

Saxenian, A. (1994) *Regional Advantage: Culture and Competition in Silicon Valley and Route 128*. Cambridge, MA: Harvard University Press.

Shannon, C. E. and Weaver, W. (1998) *The Mathematical Theory of Communication*. Urbana, IL: University of Illinois Press.

Szulanski, G. (1996) Exploring internal stickiness: impediments to the transfer of best practice within the firm. *Strategic Management Journal*, 17: 27–43.

Vernon, R. (1966) International investment and international trade in the product cycle. *Quarterly Journal of Economics*, 80: 190–207.

Vernon, R. (1979) The product cycle hypothesis in a new international environment. *Oxford Bulletin of Economics and Statistics*, 41: 255–67.

Young, S., Hood, N. and Peters, E. (1994) Multinational enterprises and regional economic development. *Regional Studies*, 28(7): 657–77.

Young, S., Hood, N. and Wilson, A. (1994) Targeting policy as a competitive strategy for European inward investment agencies. *European Urban and Regional Studies*, 1(2): 143–59.

Zander, I. (1999) How do you mean 'global'? An empirical investigation of innovation networks in the multinational corporation. *Research Policy*, 28(2–3): 195–213.

Part II

Governing External Knowledge Relations

4
Do Good Threats Make Good Neighbours? Social Dilemmas in MNC Networks

Margit Osterloh and Antoinette Weibel[1]

Introduction

Articles on strategic networks have been plenty in recent times adding to our understanding of how to establish and maintain successful interorganizational relationships. However, little attention has been paid to the detailed governing and structuring of these relations on a more operational level (Grandori, 1997; Sobrero and Schrader, 1998). Yet this is exactly the level we have to turn to if we want to explain why so many strategic networks still fail despite the obvious advantages of not going it alone.

Literature on governance structures and processes in the MNC has long been dominated by transaction cost literature (Oliver and Ebers, 1998; Williamson, 1999), which offers clear-cut devices for the choice of the right governance structure in a given situation. The elegance of this approach has been built on the assumption of opportunism. However this assumption comes at a high price in that it may firstly, make MNCs overinvest in safeguards (Madhok, 2000). Secondly, it may make certain forms of cooperation impossible (Madhok and Tallman, 1998) and thirdly, what we consider to be the most important point, it may further the intent of the partners to behave opportunistically, which can eventually lead to a spiral of distrust (Ghoshal and Moran, 1996; Osterloh and Frey, 2000).

Recently a more realistic and complex approach has emerged. Network partners are shown to have both private and common benefits from allying (Khanna, Gulati and Nohria, 1998) which ties them together in a situation, commonly referred to as 'social dilemma'. A social dilemma arises if interdependent partners face the problem that private and common benefits cannot be optimized together and the optimization of the private benefits yields outcomes leaving all partners worse off than feasible alternatives (Dawes, 1980; Ostrom, 1998). Learning alliances can be very often thought of as an example of a social dilemma situation (Kale, Singh, and Perlmutter, 2000). Partners cannot know beforehand whether their counterparts primarily seek to further their own interest, which in a race to learn would be to outlearn each other (Hamel and Doz, 1989). On the other hand the network MNCs also realize that in a situation of distrust and holding back the common interests of an interorganizational learning process cannot be advanced. The main question is: What can network partners do to overcome social dilemmas?

We would like to answer this question by discussing two approaches which have been underrated in the network literature so far: the public choice and the crowding-out theory. In a first section we show how re-framing the situation using a public choice approach can sharpen the governance problem of networks. Depicting networks as a social dilemma provides a much more adequate definition of the situation than other approaches such as transaction cost theory can offer. We then provide in a second section the solutions offered to this collective action problem in the light of public choice. This allows us also to tie different network research streams together. A third section is dedicated to conduct a more precise analysis of trust, which plays a key role in overcoming social dilemmas. Here public choice, too, dismisses the underlying dynamics of the process of trust building. We show that in certain conditions a crowding-out of trust may occur. It is therefore necessary that the crowding out theory is introduced in the network context. Finally, in a fourth section we draw the theories together to analyse empirical studies.

Strategic networks as a source of social dilemmas

MNCs enter strategic networks in order to gain market access, to reduce market uncertainty, to impose industry standards, to reap economies of scale and for a variety of other reasons (Sydow, 1992). An increasingly important reason to collaborate has been the aim to complement a MNC's capabilities i.e. to learn from each other. Strategic networks are seen as a vehicle to gain access to capabilities of other MNCs which otherwise would have been very time-consuming to build (Nooteboom,

1999). In this view strategic networks are collaborations of legally independent MNCs in order to gain competitive advantage. However, as we will discuss in the following, in the process of creating common values the network MNCs are facing a social dilemma, which *cannot* be solved by traditional means i.e. hierarchy or privatization, as in networks property rights and decision rights are always diffused, with no single MNC holding the majority of property and decision rights.

The process of interorganizational learning creates operative interdependencies. To highlight this issue we use Thompson's (1967) typologies and distinguish among three possible types. *Pooled interdependencies* exist if the network MNCs draw on the same pool of resources, e.g. common finances. *Sequential interdependencies* typically evolve in relationships where the output of one partner is at the same time the input of another partner. Finally, *reciprocal interdependencies* arise if input–output relationships switch, e.g. interorganizational development where partner MNCs change prototypes and specifications forth and back. Typically in strategic networks all types of interdependencies are present at the same time. However, reciprocal interdependencies constitute a special sort of governance problem. The required fit of interorganizational processes and operations can often not be specified *ex ante* and only gradually evolves through interaction (Borys and Jemison, 1989). In addition to the coordination problem the MNCs are facing a motivation problem: Can we trust the other MNCs to act in our interest i.e. to join the collective action?

Pooled and reciprocal interdependencies in networks always constitute a social dilemma. The analysis of social dilemmas has been strongly influenced by Garett Hardin (1968), who described a form of social dilemma, which he called 'tragedy of the commons'. The commons are community pastures where herders were free to graze their cattle. Because there usually exist no entry barriers it is rational if each individual herder adds more animals than the social optimum. This, however, leads to a social trap or even to a common disaster, i.e. the commons are overgrazed and eventually may be destroyed. In a more general sense Dawes (1980: 169) defines social dilemmas:

> Social dilemmas are defined by two simple properties: (a) each individual receives a higher payoff for a socially defecting choice (e.g. having additional children, using all the energy available, polluting his or her neighbours) than for a socially cooperative choice, no matter what the other individuals in society do, but (b) all individuals are better off if all cooperate than if all defect.

There has been a lot of research inspired by this rather bleak outlook. Whereas a lot of social scientists have been busy proving the ubiquity of social dilemmas some research traditions have tackled the problem of how individuals, MNCs and societies can deal with them. Public choice has been one domain adding fruitful insights to the problem.

In public choice two problems are identified with social dilemmas. The *first problem*, which is illustrated so vividly by Hardin's (1962) 'tragedy of the commons', is the luring temptation to *freeride* on others' efforts. Because no herdsman can be excluded from the common pasture, it is rational to become a freerider, as he will be better off whatever the others do. If the others don't rise to the occasion he can herd his cattle as well as profit from the future existence of the commons. If all herdsmen keep on increasing their share then at least for a limited time he too can reap the profits of a better fed herd (Stroebe and Frey, 1982). The *second problem* poses a second-order dilemma. Rational individuals would indeed understand the situation and try to find ways to avoid the overgrazing of the commons. However, finding ways, establishing rules and norms of behaviour and monitoring the compliance are actions which again encourage freeriding. Elster (1989: 41) explains it as following:

> Punishment almost invariably is costly to the punisher, while the benefits from punishment are diffusely distributed over all members. It is, in fact, a public good.

In learning networks we can find the most obvious accounts for social dilemmas. MNCs can never be sure whether their partners are out to get them i.e. to steal their competencies, or whether their partners would like to initiate an interorganizational learning process for the sake of common learning. The literature provides accounts of networks which have fallen into the social trap e.g. the case of the joint venture between the French Thompson and the Japanese JVC (Doz and Hamel, 1998). Entering the network seemed to be a win–win situation for both partners. JVC was granted entry to the European market and saw the possibility of defining the industry standard and Thompson wanted to use the process know-how of JVC. Yet Thompson entered the network with a hidden agenda. Its aim was to match JVC's competencies in a few years time. Whilst at first JVC did not consider Thompson to be a threat and rather openly shared its knowledge in the common production site they were soon to discover that Thompson had indeed learned a lot and came dangerously close in filling the knowledge gap. By this time it was already too late to alter the conditions because JVC had become rather dependent on Thompson.

On the other hand, examples of strategic networks thriving despite the obvious dilemma situation also abound, starting with Japanese supplier–producer relationships (Dyer, 1997), multilateral learning networks such as the network initiated by the airport corporation of Frankfurt, the FAG (Duschek, 1998) and virtual or dynamic strategic networks such as the Australian TCG (Miles and Snow, 1994). Mathews (1993) describes this interesting network as 'a cluster of autonomous MNCs, numbering approximately 24, each of which specializes in a particular facet of information technology services or product development'. TCG has developed a set of internal rules governing the collaboration and preventing the single MNCs from falling into the social trap. The MNCs seem to have solved the social dilemma and reap the advantages of an interorganizational development process called 'triangulation strategy'. The essence of this is to develop new ideas with two outside partners, one major customer and one technical partner with needed competencies, and to involve other TCG MNCs in the process (Miles *et al.*, 1997).

What, then, is it that certain MNCs know in order to avoid or to solve the social dilemmas? We will turn to this question in the next section.

Solutions to social dilemmas

The grim predictions evoked by the analysis of dilemma situations ran counter to many everyday experiences and are also challenged by a considerable amount of empirical evidence in the dilemma research. On a very general level there are two routes to solve the problems of cooperation. The first and traditional solution is to establish an overarching authority, an outside force to induce others to do things for the common benefits. This is what Hardin (1968: 314) recommended when he made the problem popular:

> if ruin is to be avoided in a crowded world, people must be responsive to a coercive force outside their individual psyches, a 'Leviathan' to use Hobbes's term.

In addition, experiments and empirical studies have shown that individuals or MNCs are capable of solving the dilemma situation by reshaping the structure of the game (Parkhe, 1993), which is the second solution. Again, there are various ways to alter the situation. On the one hand they can diminish the incentives to freeride. On the other they can build up trust. Thereby their credibility is strengthened, which eases collaboration.

We will now discuss these solutions in a network context.

Establishing authority

In traditional transaction cost economics (TCE) theory, MNCs cope with rising interdependencies in networks, i.e. high asset specificity, by turning to a hierarchical structure. Hybrid structures must ultimately fail to coordinate highly interdependent activities in dynamic markets as the necessary bilateral (or multilateral) consent takes time and is conducive to misunderstandings (Williamson, 1996). According to TCE, through integrating further adaptations can be devised by fiat. Also in an integrated MNC competition between the units is lower than it is usually between legally and economically independent network MNCs which softens the underlying dilemma as well.

Yet in reality networks do exist in situations of high asset specificity and dynamic markets, a fact not easily accounted for in traditional transaction cost theory (Holmström and Roberts, 1998). One possible explanation is that MNCs have found ways to lower long-term transaction costs and to counter the social dilemma in networks by employing hierarchical coordination devices such as quality controls or network standards next to other governance structures. Alchian and Demsetz (1972) have given a theoretical explanation of why actors voluntarily choose to set up a hierarchy. In their view the social dilemma of team production can only be overcome if a central agent exists. Her or his interest to further the common good is secured as s/he is entitled to the residual gains. However, their considerations rest on the assumption that the central agent owns the residual rights. This is of course not the case in networks where no network MNC owns all property rights. The sanctioning power of the focal MNC is therefore lower than the sanctioning power of the central agent in the model of Alchian and Demsetz (1972). Empirically it is well established that one or two MNCs often take the strategic lead in networks and act as hub MNCs to control their partners (Rugman and D'Cruz, 2000; Sydow, 1992). Sobrero and Schrader (1998: 587) describe different means of formal hierarchical coordination, such as establishing a command structure, incentives system and operating procedures to be employed by the hubs.

But these solutions do not work well in the situation of information asymmetry (Miller, 1992). A problem, especially virulent in learning networks, is that interorganizational learning processes are often of tacit nature. Firstly, the hub MNC encounters problems in choosing the right partner as they can easily hide their real value and their 'true' intention. Secondly, even in an ongoing cooperation it is very difficult to understand the partner's competencies. Finally, it is virtually impossible to judge *ex-post* what the contributions of the single MNCs were to the network performance. On the one hand it is very difficult to untangle the influence of the general market dynamics, luck and efforts of the MNCs on the network performance. On the other hand as network cooperation may lead to leverage effects who can tell the effect of a single contribution to this special surplus?

The partial failure of the hierarchical solution leads us to the second possibility: deliberate strategies to change the structure of the game.

Lower incentives to freeride

Recently network literature has drawn on game theory to offer further solutions to the network dilemma. Parkhe (1993: 797) emphasizes:

> However, the dilemma's relentless logic and the inherent instability introduced into the relationship by each partner's uncertainty regarding the other's next move may be responsive to deliberate strategies that do not necessarily accept circumstances as given, but rather seek to reshape the alliance structure to create the conditions for robust cooperation.

From a game theoretic point of view social dilemmas can be transformed into different games that make cooperation more plausible.

It is by now a well-established fact that a *long shadow of the future* makes cooperation more probable. Once there is potential for a long cooperation with no clear-cut deadline, strategies of reciprocity are encouraged. The network partners can commit themselves to punish non-cooperators sufficiently such as in the famous 'tit for tat' strategy posited by Axelrod (1984). 'Tit for tat' refers to thin forms of reciprocity that can be understood as a pattern of exchange i.e. a mutually contingent exchange of benefits or gratifications. By contrast, thick forms of reciprocity rely on the idea of reciprocity as a norm in the sense that 'you should give benefits to those who give you benefits' (Gouldner, 1960).

The network MNCs can themselves prolong the shadow by investing in the relationship. High set-up costs for structuring the further relation, e.g. creating interorganizational information systems, serve both for lowering ongoing transaction costs *and* creating a longer time horizon (Dyer, 1997).

However, to resolve social dilemmas solely by depending on these thin reciprocity norms is somewhat problematic. Tit for tat strategies can only evolve if the termination of the relationship is unknown (Fudenberg and Maskin, 1986), which is most often the case in strategic networks. The level of cooperation can still not be specified *ex ante*. In other words, there is still plenty of room to freeride although there will be no complete freeriding any longer. This is because once the uncertainty of the duration is introduced the number of possible equilibria explode (Abreau, 1988). In other words there is no guarantee how much of the private interests will be subjected to common interests. A higher degree of cooperation can only be achieved if the network MNCs can credibly communicate their trustworthiness i.e. their dedication to act in the common interest. A thicker form of reciprocity than promoted in game theoretic approaches is needed.

Trust building

In public choice trust is understood as 'lubricant of the social system' (Arrow, 1974: 23), i.e. a very efficient way out of the social dilemma. Also in many network theories trust has been found to be the glue that keeps the partners together (Bradach and Eccles, 1989; Lane and Bachmann, 1998; Nooteboom, 1996; Zajac and Olsen, 1993). Many advantages are attributed to trust: trust encourages the disclosure of truthful information (Ring and Van de Ven, 1992), lowers monitoring costs (Barney and Hansen, 1994) and, most importantly in the context of the social dilemma, 'trust promises a potentially high degree of effort exertion not only in routine tasks but in dynamic responses to new situations' (Sako, 1992: 48), i.e. there will be less freeriding. Although the literature on trust is burgeoning there is still a lot of confusion attached to the concept. Therefore we need to be precise about what we mean by trust.

To trust somebody means to act *as if* there is no doubt that the other will act in our interest. A trustor allows herself or himself to be vulnerable towards the trustee because s/he believes that the trustee is intrinsically motivated to reciprocate. We thereby explicitly rule out a definition of trust most economists use, i.e. trust in the sense of calculative trust. In this aspect we agree with Williamson (1996: 261) that 'calculative relations are best described in calculative terms'. Trust is not understood as a subclass of rational decision-making under uncertainty/risk. We go along with Luhmann (1989) and propose that people who are trusting act as if there was no risk precisely because the other person is seen as trustworthy. Mayer, Davis and Schoorman (1995) have defined trust as the willingness of a party to be vulnerable to the action of another party based on the expectation that the other will perform a particular action important to the trustor, irrespective of the ability to monitor or control that other party.

The trustee is intrinsically motivated to cooperate because s/he either likes the trustor and/or because s/he adheres to the norm of reciprocity in believing that this is the right thing to do. Elster (1989: 192) has coined this adherence as 'everyday Kantianism' which says 'that one should cooperate if and only if universal cooperation is better for everybody than universal defection'.

By choosing this definition we wish to highlight two important dimensions: firstly, trust is not understood as 'rational decision under uncertainty'. Unlike most economists, we believe it is very important to separate trust from rational calculus. A trustor does not rely on thin norms of reciprocity as documented by the tit for tat strategy. Rather to trust means to expect the partner to reciprocate in a more general sense, i.e. to act in one's interest even if her/his interest is compromised *and* even if it eventually will not pay out fully. It is the expectation that the other person also relies on these thicker forms of reciprocity.

Secondly, trust takes time to build. It is never 'just there' to act as 'lubricant of transactions'. We do believe that it is on the contrary very important to understand the process of building trust, which is messy and uncertain to say the least. Also in the early phases of the process monitoring, incentives, and norms of thin reciprocity play a pivotal role. Yet the ideal mix of these three governing devices, i.e. authority, incentives and trust, has not been established so far. In the next section we will stress the danger of provoking a spiral of distrust if the network partners rely too much on authority, incentives and thin norms of reciprocity. We will show the conditions in which this can happen.

No dilemma with the 'right' mix of governance devices

Let us first consider the creation of trust in strategic networks. As has been made clear, the development of trust cannot be taken for granted in a social dilemma situation where self-interest and temptations to act opportunistically exist. Moreover, there is no way to distinguish trustworthy partners from others *ex ante* due to the shown information asymmetries. In such a situation trust can only gradually evolve in the network.

From what we know from empirical studies of the development of strategic networks (Doz, Olk and Ring, 2000; Larson, 1992; Ring and Van de Ven, 1994; Zajac and Olsen, 1993) we can sketch a very simple model of trust building. Trust evolves only gradually in a trial and error process. Information on how the network partners perform and cooperate are fed back and serve to re-evaluate the trustworthiness of the partners in every phase (Mayer, Davis and Schoorman, 1995). Especially in these early phases emotional ties are weak or absent. Therefore monitoring not only protects the network partners but also uncovers valuable information (Lewicki and Bunker, 1995).

In this situation MNCs have to be particularly cautious because monitoring has the potential to further trust *but* it has also the potential to destroy trust. MNCs have to consider the conditions of when such a crowding-out effect will take place. We therefore introduce the crowding-out theory carefully to delineate these conditions (Frey, 1997; Osterloh and Frey, 2000).

Analysis of the crowding effect

There exists overwhelming theoretical and empirical evidence that intrinsic and extrinsic motivation are not additive. Rather, there is a systematic dynamic relationship between the two. This dependence has been shown to exist in a large number of careful experiments undertaken by Deci and his group (Deci, 1971; Deci, Koestner and Ryan, 1999). These relationships between intrinsic and extrinsic motivation are called *crowding effects* (Frey, 1997). These effects make both kinds of motivation *endogeneous* variables. The *crowding-out effect* posits a negative relationship between intrinsic and extrinsic motivation. When external incentives – rewards or controls – are perceived to be controlling by the member MNC affected, intrinsic motivation tends to be undermined.

Theoretically the crowding effect is based on *cognitive evaluation theory* (Deci, 1985) and on the economic theory of motivation (Frey, 1997). Further the *theory of organizational justice* (Lind and Tyler, 1988) and *psychological contract theory* (Rousseau, 1995) have added to our understanding. Taken together they specify the conditions under which intrinsic motivation (to act trustworthily) is decreased or increased.

According to cognitive evaluation theory intrinsic motivation depends on the perceived locus of causality, which shifts if the drive to act is attributed to an external influence (Deci and Flaste, 1995). The actor considers the person undertaking the outside intervention to be responsible. The important finding is that the shift in the locus of causality does not always take place. Each internal intervention has two aspects: (1) the *controlling aspect* strengthens the perceived external control and the feeling of being directed from the outside. (2) The *informing aspect* adds to one's perceived competence and thereby strengthens the feeling of internal control. When external incentives – rewards or commands – are perceived to be controlling intrinsic motivation tends to be undermined. The overall effect depends on which influence is perceived to be stronger.

Organizational justice literature characterizes the conditions when such a shifting of the locus of causality is more likely. It emphasizes that it is especially important how extrinsic benefits are applied if negative effects are to be avoided. Firstly, a fair distribution of revenues is shown to be very important. Adams (1963) was the most influential in pointing out that the allocation of benefits and costs within a group should be equitable. However, the equity criterion can only be applied if contributions and outcomes can be measured effectively which is not the case in learning network dilemmas. In such a case equality may be the more appropriate principle. According to Grandori and Neri (1999: 50), applying an equality norm helps to save information costs and to foster a climate of harmony and trust. Secondly, in recent time the focus has been laid on procedural justice. People place importance not only on what they get but also on how their share was decided and distributed. Empirically, procedural justice is shown to be especially important if the extrinsic benefits are not considered equitable. In such a case fair procedures enhance the satisfaction with the negative results (Lind and Tyler, 1988). Nevertheless procedural justice has a great impact on the acceptance of existing authorities. Whether or not rules, standard operating procedures, controls and other hierarchical devices designed to solve the social dilemma are accepted is dependent on legitimacy or authority. Tyler and Degoey (1996: 493) have shown that views of the legitimacy of an authority are almost only influenced by the procedural fairness administered by that authority. Finally, there is a growing body of empirical studies that directly show the influence of fair processes on trust (Alexander and Rudermann, 1987; Koorsgard, Schweiger and Sapienza, 1995).

Psychological contract theory adds to our understanding of the dynamics of crowding effects or more directly of the dynamics of trust and distrust evolution. First of all a breach of the individual beliefs about the terms and conditions of a reciprocal exchange, i.e. the breach of the psychological contract, can lead to a loss of trust (Robinson, 1996). Robinson (1996: 578) has established empirically that a psychological breach 'undermines two conditions leading to trust – judgment of integrity and beliefs in benevolence – that in turn reduce employees' contributions'. As a consequence the individual's contributions will be reduced and the relations are reduced to a purely transactional exchange. Thereby a spiral of distrust evolves. However, Robinson (1996: 590) also established that higher trust at the outset leads to a lower decline in trust following the psychological contract breach. Trusting individuals are less likely to perceive a breach. In such a way trust begets trust not only by influencing the trusting behaviour but also by influencing each other's perceptions.

Balancing extrinsic and intrinsic governance devices

As a consequence the design of governance structures in networks has to take crowding effects into account. As far as we can tell from the theoretical analysis four aspects should be taken into consideration to further the intrinsic obligation to behave trustworthly and to avoid crowding-out (Osterloh and Frey, 2000):

1. **Participation** is a key feature in avoiding crowding effects. Put differently, unilateral commands may trigger a vicious circle, i.e. they diminish self-initiatives and produce the kind of conduct which they were meant to avoid (McGregor, 1960). Contrary to that participation, firstly, avoids the shift of the locus of causality as it raises the perceived self-determination of the network MNCs and thereby strengthens intrinsic motivation to behave trustworthily. Secondly, participation is a very important criterion of fair procedures. By now it has been well established that procedures giving members a voice in the decision-making process tend to enhance the acceptance of even unfavourable decisions (Greenberg, 1987). Thirdly, the common design of governance mechanisms, i.e. participation in creating a design structure, increases the likelihood of shared perceptions in the underlying psychological contracts.
2. **Self-organizing principles** produce willingness to obligate social norms for their own sake. In this sense, they can foster the coordination of divergent interests of different network members (Ostrom, 1990).
3. **Personal relationship** and **communication** is a precondition for establishing and maintaining psychological contracts of relational nature. As experimental evidence shows, personal relationship strongly raises the intrinsic motivation to cooperate (Dawes, van de Kragt and Orbell, 1988; Frey and Bohnet, 1995). Whereas communication is closely related to minimize possible incongruences between the perception of different agents thereby reducing the likelihood of a perceived breach (Morrison and Robinson, 1997). Interorganizational working groups, transfer of personnel and

substantial information sharing are seeds to relational exchanges and can be observed in longstanding strategic networks (Dyer, 1997; Sako, 1992).

4. **Fair procedures** add legitimacy to network authorities, may curb dissatisfaction with inequitable extrinsic rewards and directly influence mutual trust. Procedures are understood to be fair if they rely on accurate information, provide consistency over time and persons, are correctable and represent the concerns of all recipients (Greenberg, 1990).

Taking these crowding effects into account ensures that governance devices act additively to combat the social dilemma. Whilst we have shown theoretical and empirical evidence on important features of the crowding effect the right mixture of the devices can only be established empirically. We therefore turn in the next section to empirical studies, which shed light on the remaining questions.

Empirical evidence

Research on networks has only just begun to study the adequate mix of governance structures in dilemma situations. In public choice on the other hand there is an impressive number of studies exploring how some communities have managed to solve severe social dilemmas over an extended period of time. The social dilemmas studied arise due to common pool resources, which because of their collective good characteristics invite freeriding of the participants.

Elinor Ostrom (1990) and her research team have been comparing successful communities worldwide and have distilled eight common governance principles. The subject of their studies are field settings in which (1) the government mechanisms are devised, applied and monitored by the players themselves and (2) the dilemma has been avoided for a long time. Her empirical work is comparable to the situation in strategic networks because in neither setting can outside force provide the solution. The network MNCs as well as the community members have to bargain for the possible solutions. However there is also one obvious difference: Ostrom deals only with tangible common pool resources whereas in strategic networks intangible collective goods are often the main problem. We will discuss the implications of this important difference further on.

We will now analyse these five empirically grounded principles which are comparable to the network context on the background of the theories we introduced. Especially we are going to show, that the structures are carefully designed to avoid crowding effects.

1. **Clearly defined boundaries.** All communities have established common property rights and boundaries of the common pool resources. Often it is made very difficult for outsiders to enter the club of pool owners and the rules of membership are clearly delineated. For example, in Törbel, a small village in Switzerland, access to the common pastures is limited to citizens (Ostrom, 1990: 62).
2. **Congruence between appropriation and provision**. The appropriation rules reflect the provision of resources and efforts. Also they are geared towards the specificities of the pool resource in question. In Törbel, for example, the cheese produced by the communities and the number of cows a family sends to the pasture determines herdsmen. However, the maintenance work such as building and maintaining roads is also delegated in proportion to the sent cattle (Ostrom, 1990: 62 f.).
3. **Rules are commonly devised.** Most individuals affected by the rules can participate in the design and modification process. In Törbel the association that commonly owns the common pool resource has annual meetings to discuss rules and policies (Ostrom, 1990: 62 f.).

These first three principles enable the community members to act. Accepted rules can be found and modified accordingly. More importantly, fairness conditions are secured. *Distributive justice* is secured by the fact that provision and appropriation are closely linked. *Procedural justice* too is held high. The common devising of rules at once secures the representation of all individuals who are affected and enables a modification of hitherto incorrect procedures.

Looking at empirical network studies one is quick to spot similarities. Access is often highly restricted and some form of participative rule-making is always present. Yet one important difference lies in the characteristic of common pool resources in strategic networks, i.e. their inherent intangible nature. Examples of *intangible* common pool resources within networks are mutual commitment, common interorganizational rules and routines, and accumulated network-specific knowledge. As we have already pointed out, the equity criterion can only be applied if contributions and outcomes can be measured effectively, which is not the case with intangible common pool resources. In such a case equality may be the more appropriate principle and will probably be more often found in longlasting networks.

The established rules have to be monitored and incentive systems have to be designed that support rule appliance:

1. **Monitoring**. The common pool resource owners often carry out monitoring themselves. This has a number of advantages: (1) monitoring provides the guardian with the added benefit of learning on the others' trustworthiness and (2) very often the guardian may keep the fine imposed. Lastly (3) only an insider, i.e. a community member, understands the context of rule breaking. By this means overreactions which otherwise could lead to a spiral of distrust can be avoided.
2. **Graduated sanctions**. Again, as long as the community members do the monitoring the context and situation is better understood. That is why they can better judge the severeness of the rule infraction and particular circumstances surrounding the deed. Sanctions can be geared to the specific circumstances. For example, individuals breaking the rules for the first time will only have to pay modest sanctions; only escalating practices of rule infraction will be punished harshly.

As shown, hierarchical governance structures are an important device to solve the social dilemma. However, according to the empirical results of Ostrom (1990) and to the theories discussed in the previous section, a crowding effect by unilateral commands has to be avoided. In the case of these successful common pool resources, managers' participation, self-organizing principles and information play key roles. Firstly, monitoring and sanctioning rules are commonly devised. Secondly, monitoring and sanctioning is delegated to the community members. No external force can be held responsible, which strengthens the feeling of responsibility. Finally, the informing aspect of monitoring seems to be very important. It is understood by the community members that they can learn about each other's trustworthiness in the monitoring process. It is also seen that sanctions can have an informing aspect if they are graduated, i.e. geared to the specific situation.

Again we have to remember that strategic networks often centre around *intangible* common pool resources. As a consequence monitoring will be much more costly. Intangible common pool resources are extremely hard to codify because they are based to a great extent on tacit knowledge. They are non-observable and non-verifiable because of their credence qualities. Credence qualities of a resource are those which, although worthwhile, cannot be evaluated in normal use. It is nearly impossible (1) to calculate what percentage these common pool resources contribute to the offered products and services of an MNC, and (2) to single out individual, highly idiosyncratic contributions to these joint efforts. Because external monitoring will be very subjective peer monitoring will be even more important. The appropriate sanctioning too is more difficult to devise. Probably the penalties will also increase with the severeness of the rule infraction and/or the frequency of the rule breaking. However, the process will evolve much more slowly.

The empirical results discussed are important first hints on how networks in the absence of a central authority can establish a mix of governance structures to solve the inherent social dilemma. They show under what conditions thick forms of reciprocity, i.e. trust, may be furthered rather than destroyed by control and sanctioning devices. These empirical results may serve as a common ground in studying knowledge-based networks, where trust will play a crucial role.

Conclusion

We have shown that success of strategic networks depends on how well network MNCs manage the underlying social dilemma of their cooperation. Hierarchical devices, thin forms of reciprocity and trust based on thick reciprocity are pivotal stepping stones in a successful path to manage the social dilemma. However, the combination of these devices is tricky, as underlying crowding effects have to be considered. Ultimately, the right mix of governance structures has to be established empirically.

In studying existing empirical work on the successful management of common pool resources important insights can be gained. In the absence of a central authority, community members have been shown to be able to create a complex set of hierarchical devices, incentives and sanctioning systems without destroying the formation of trust between the members of the community. However, we expect the management of social dilemmas in strategic networks compared to common pool resources to be even more complex.

The common value-creation in strategic networks leads to a more severe social dilemma than in the studied communities because of the heightened measurability problem. On the one hand a MNC's individual input to the common good, i.e. collective learning, cannot be measured due to synergy effects. On the other hand the output of the cooperation efforts cannot be measured as collective learning typically shows credence qualities. Trust will therefore play a more important role in networks than in the studied communities.

The right mix of governance structures in networks can only be established by further empirical research based on the delineated framework by which we have offered a precise terminology to understand and analyse governance issues of social dilemmas in strategic networks.

Note

1. *Acknowledgement*: We are very grateful to Sandra Rota for critical comments and to Daniel Bastian for basing a lecture on the essay. Also we learned a lot from the helpful reviews at the LINK conference in Copenhagen – special thanks to Ingmar Bjorkman and Ulf Holm. We also like to thank Margaret Levi whose (now published) working paper inspired us to the title of the essay.

References

Abreau, P. (1988) On the theory of infinitely repeated games with discounting. *Econometrica*, 80(4): 383–96.

Adams, J. S. (1963) Toward an understanding of inequity. *Journal of Abnormal and Social Psychology*, 67: 422–63.

Alchian, A. and Demetz, H. (1972) Production, information costs, and economic organization. In A. Alchian (1977) *Economic Forces at Work*. Indianapolis: Liberty Press.

Alexander, S. and Rudermann, M. (1987) The role of procedural and distributive justice in organizational behavior. *Social Justice Research*, 1(2): 177–98.

Arrow, K. J. (1974) Gifts and exchanges. In Edmund S. Phelps (ed.), *Altruism, Morality and Economic Theory*. New York: Russell Sage Foundation.

Axelrod, R. (1984) *The Evolution of Cooperation*. New York: Basic Books.

Barney, J. B. and Hansen, M. H. (1994) Trustworthiness as a source of competitive advantage. *Strategic Management Journal*, 15(Winter, special issue): 175–90.

Borys, B. and Jemison, D. B. (1989) Hybrid arrangements as strategic alliances: theoretical issues in organizational combinations. *Academy of Management Review*, 14(2): 234–49.

Bradach, J. L. and Eccles, R. G. (1989) Price, authority, and trust: from ideal types to plural forms. *Annual Sociological Review*, 15: 97–118.

Dawes, R. M. (1980) Social dilemmas. *Annual Review of Psychology*, 31: 169–93.

Dawes, R. M., van de Kragt, A. J. C. and Orbell, J. M. (1988) Not Me or Thee but We: the importance of group identity in eliciting cooperating in dilemma situations – experimental manipulation. *Acta Psychologica*, 68: 83–97.

Deci, E. L. (1971) Effects of externally mediated rewards on intrinsic motivation. *Journal of Personality and Social Psychology*, 22: 113–20.

Deci, E. L. (1985) *Intrinsic Motivation and Self-Determination in Human Behavior*. New York: Plenum Press.

Deci, E. L. and Flaste, R. (1995) *Why We Do, What We Do. Understanding Self-Motivation*. Rochester: Penguin Books.

Deci, E. L., Koestner, R. and Ryan, R. M. (1999) The undermining effect is a reality after all – extrinsic rewards, task interest, and self-determination: reply to Eisenberger, Pierce and Cameron (1999) and Lepper, Henderlong and Gingras (1999). *Psychological Bulletin*, 125: 692–700.

Doz, Y. L. and Hamel, G. (1998) *Alliance Advantage: The Art of Creating Value through Partnering*. Boston, MA: Harvard Business School Press.

Doz, Y., Olk, P. M. and Ring, P. S. (2000) Formation processes of R&D consortia: Which path to take? Where does it lead? *Strategic Management Journal*, 21(March): 239–66.

Duschek, S. (1998) Kooperative Kernkompetenzen – Zum Management einzigartiger Netzwerkressourcen. *Zeitschrift Führung und Organisation*, 4: 230–6.

Dyer, J. H. (1997) Effective interfirm collaboration: how firms minimize transaction costs and maximize transaction value. *Strategic Management Journal*, 18(7): 535–56.

Elster, J. (1989) *The Cement of Society*. Cambridge, New York, Victoria: Cambridge University Press.

Frey, B. S. (1997) *Not Just for the Money: An Economic Theory of Personal Motivation*. Cheltenham, UK and Brookfield, USA: Edward Elgar.

Frey, B. S. and Bohnet, I. (1995) Institutions affect fairness: experimental investigations. *Journal of Institutional Theoretical Economics*, 151(2): 286–303.

Fudenberg, D. and Maskin, E. (1986) The folk theorem in repeated games with discounting or with incomplete information. *Econometrica*, 54(3): 533–44.

Ghoshal, S. M. and Moran, P. (1996) Bad for practice: a critique of the transaction cost theory. *Academy of Management Review*, 21: 13–47.

Gouldner, A. W. (1960) The norm of reciprocity: a preliminary statement. *American Sociological Review*, 25(2): 161–78.

Grandori, A. (1997) An organizational assessment of interfirm coordination modes. *Organization Studies*, 18(6): 897–925.

Grandori, A. and Neri, M. (1999) The fairness properties of interfirm networks. In Anna Grandori (ed.) *Interfirm Networks: Organization and Industrial Competitiveness*. London and New York: Routledge.

Greenberg, J. (1987) A taxonomy of organizational justice theories. *Academy of Management Review*, 12(1): 9–22.

Greenberg, J. (1990) Organizational justice: yesterday, today, and tomorrow. *Journal of Management*, 16(2): 399–432.

Hamel, G. and Doz, Y. L. (1989) Collaborate with your competitors and win. *Harvard Business Review*: 133–9.

Hardin, G. (1968) The tragedy of the commons. *Science* (162): 1243–8.

Holmström, B. and Roberts, J. (1998) The boundaries of the firm revisited. *Journal of Economic Perspectives*, 12(4): 73–94.

Kale, P., Singh, H. and Perlmutter, H. (2000) Learning and protection of proprietary assets in strategic alliances: building relational capital. *Strategic Management Journal*, 21(March): 217–37.

Khanna, T., Gulati, R. and Nohria, N. (1998) The dynamics of learning alliances: competition, cooperation, and relative scope. *Strategic Management Journal*, 19: 193–210.

Koorsgard, M. A., Schweiger, D. M. and Sapienza, H. J. (1995) Building commitment, attachment, and trust in strategic decision-making teams: the role of procedural justice. *Academy of Management Journal*, 38(1): 60–84.

Lane, C. and Bachmann, R. (1998) *Trust Within and Between Organizations: Conceptual Issues and Empirical Applications*. Oxford and New York: Oxford University Press.

Larson, A. (1992) Network dyads in entrepreneurial settings: a study of the governance of exchange relationships. *Administrative Science Quarterly*, 37: 76–104.

Lewicki, R. J. and Bunker, B. B. (1995) Trust in relationships: a model of trust development and decline. In Barbara B. Bunker and Jeffrey Z. Rubin (eds) *Conflict, Cooperation, and Justice*. San Francisco: Jossey-Bass.

Lind, E. A. and Tyler, T. R. (1988) *The Social Psychology of Procedural Justice*. New York: Plenum.

Luhmann, N. (1989) *Vertrauen. Ein Mechanismus der Reduktion sozialer Komplexität*. Stuttgart: Enke.

Madhok, A. (2000) Transaction (in-)efficiency, value (in-)efficiency and inter-firm collaboration. In David O. Faulkner and Mark de Rond (eds) *Cooperative Strategy. Economic, Business, and Organizational Issues*. Oxford and New York: Oxford University Press.

Madhok, A. and Tallman, S. B. (1998) Resources, transactions, and rents: managing value through inter-firm collaborative relationships. *Organization Science*, 9(3): 326–39.

Mathews, J. (1993) TCG R&D networks: the triangulation strategy. *Journal of Industry Studies*, 1(1): 65–74.

Mayer, R. C., Davis, J. H. and Schoorman, F. D. (1995) An integrative model of organizational trust. *Academy of Management Review*, 20(3): 709–34.

McGregor, D. (1960) *The Human Side of Enterprise*. New York: McGraw-Hill.

Miles, R. E. and Snow, C. C. (1994) *Fit, Failure, and the Hall of Fame: How Companies Succed or Fail*. New York: Free Press.

Miles, R. E., Snow, C. C., Mathews, J. A., Miles, G. and Coleman, H. J. (1997) Organizing in the knowledge age: anticipating the cellular form. *Academy of Management Executive*, 11(4): 7–20.

Miller, G. (1992) *Managerial Dilemmas: The Political Economy of Hierarchy*. Cambridge: Cambridge University Press.

Morrison, E. W. and Robinson, S. L. (1997) When employees feel betrayed: a model of how psychological contract violation develops. *Academy of Management Review*, 22(1): 226–56.

Nooteboom, B. (1996) Trust, opportunism and governance: a process and control model. *Organization Studies*, 17(6): 985–1010.

Nooteboom, B. (1999) *Inter-Firm Alliances*. London and New York: Routledge.

Oliver, A. L. and Ebers, M. (1998) Networking network studies: an analysis of conceptual configurations in the study of inter-organizational relationships. *Organization Studies*, 19(4): 549–83.

Osterloh, M. and Frey, B. S. (2000) Motivation, knowledge transfer, and organizational forms. *Organization Science*, 11(5): 538–50.

Ostrom, E. (1990) *Governing the Commons: The Evolution of Institutions for Collective Action*. Cambrige, New York, Victoria: Cambridge University Press.

Ostrom, E. (1998) A behavioral approach to the rational-choice theory of collective action. *American Political Science Review*, 92(1): 1–22.

Parkhe, A. (1993) Strategic alliance structuring: a game theoretic and transaction cost examination of interfirm cooperation. *Academy of Management Journal*, 36(4): 794–829.

Ring, P. and Van de Ven, A. H. (1992) Structuring cooperative relationships between organizations. *Strategic Management Journal*, 13: 483–98.

Ring, P. and Van de Ven, A. H. (1994) Developmental processes of cooperative interorganizational relationships. *Academy of Management Review*, 19(1): 90–118.

Robinson, S. L. (1996) Trust and breach of the psychological contract. *Administrative Science Quarterly*, 41: 574–99.

Rousseau, D. M. (1995) *Psychological Contracts in Organizations: Understanding Written and Unwritten Agreements*. Thousand Oaks, CA: Sage.

Rugman, A. and D'Cruz, J. (2000) The theory of the flagship Firm. In David O. Faulkner and Mark de Rond (eds) *Cooperative Strategy: Economic, Business and Organizational Issues*. Oxford and New York: Oxford University Press.

Sako, M. (1992) *Prices, Quality and Trust*. Cambridge: Cambridge University Press.

Sobrero, M. and Schrader, S. (1998) Structuring inter-firm relationships: a meta-analytic approach. *Organization Studies*, 19(4): 585–615.

Stroebe, W. and Frey, B. S. (1982) Self-interest and collective action: the economics and psychology of public goods. *British Journal of Social Psychology*, 21: 121–37.

Sydow, J. (1992) *Strategische Netzwerke: Evolution und Organisation*. Wiesbaden: Gabler.

Thompson, J. D. (1967) *Organizations in Action*. New York: McGraw-Hill.

Tyler, T. R. and Degoey, P. (1996) Collective restraint in social dilemmas: procedural justice and social identification effects on support for authorities. *Journal Pers. Soc. Psychol.*, 69: 482–97.

Williamson, O. E. (1996) Comparative economic organization: the analysis of discrete structural alternatives. In Oliver E. Williamson (ed.) *The Mechanisms of Governance*. Oxford: Oxford University Press.

Williamson, O. E. (1999) Strategy research: governance and competence perspectives. *Strategic Management Journal*, 20: 1087–108.

Zajac, E. J. and Olsen, C. P. (1993) From transaction cost to transactional value analysis: implications for the study of interorganizational strategies. *Journal of Management Studies*, 30: 131–45.

5
Learning across Borders: Organizational Learning and International Alliances

Marjorie A. Lyles and Charles Dhanaraj[1]

Introduction

This chapter presents a conceptual framework of organizational learning that integrates process models of learning (Argyris and Schon, 1974; Lyles, 1988; Nonaka, 1994; Ring and Van de Van, 1992, 1994) with the structural models of learning (Badaracco, 1991; Mowery *et al.*, 1996). We apply this framework by using it to analyse two published case studies on organizational learning in international joint ventures (IJVs).

Research on learning in alliances has two merits. First, it deals with a critical dimension of an alliance, which is central to its success. Second, it also provides a unique empirical context for learning research. For example, Lyles (1988) examined how organizations learn from their experiences in joint venturing, and if the firm had any cumulative learning from its previous alliances or critical knowledge factors that influenced the learning processes. In another paper, Lyles and Salk (1996) focused on how Hungarian alliances learned from their foreign parents, addressing the relationship between organizational factors, learning and performance.

Strategy research has shown that tangible resources such as capital, land and labour, which were the basis of neo-classical economics and industrial organization, were insufficient to explain firm performance and has suggested knowledge as a valuable resource and a key determinant of firm performance (Nelson and Winter, 1982; Winter, 1987; Hedlund, 1994). Interorganizational collaboration in the form of alliances has proliferated because of its potential to provide new sources of knowledge creation, and such linkages have become critical to knowledge diffusion and technology development (Powell *et al.*, 1996; Powell, 1998).

Given this duality of the research focus, we attempt to develop a learning framework that will inform both the fields – international alliances as well as organizational learning. Fundamentally, an alliance is an organizational form and as such, a generalized theory of organizational learning should apply equally to an alliance. Also, given the sharp focus of learning in an international alliance, where organizations from more than one country come together to share complementary knowledge either to create new knowledge or exploit existing knowledge, this provides a rich empirical context for testing a generalized theory of organizational learning. Our essay builds on Lyles' (1988) organizational learning model and expands the learning processes identified earlier, as well as recent advances in the theory of social capital and knowledge-based theory of the firm.

Following a brief overview of the literature in this area, we attempt to provide some broad opportunities and necessity for a unified theory of organizational learning. Then we present our proposed theoretical framework, first presenting the process issues, and then the structural issues. We provide two case studies from previously published sources and use these to illustrate our proposed theoretical framework. We close by discussing the implications of our framework and its potential for research and practice.

Organizational learning and alliances: research overview

The realization that knowledge is a strategic asset for a firm (Winter, 1987; Grant, 1996) has been fundamental to the progress of research on organizational learning. Knowledge is organizationally embedded, and often the most valuable component of knowledge is tacit and not explicit (Nelson and Winter, 1982; Badaracco, 1991; Dosi, 1988; Hedlund, 1994; Polanyi, 1966). This makes organizational learning more complex,. and acquiring new knowledge necessitates absorptive capacity (Cohen and Levinthal, 1990). Traditional assumptions that knowledge can flow freely within a firm have been questioned, and increasingly we realize that knowledge transfer is sticky even within a firm (Szulanski, 1996; Simonin, 1997, 1999). Also the socialization of problems and solutions is significant in organizational knowledge creation (Nonaka, 1994; Nonaka and Takeuchi, 1995).

Organizations can create new knowledge by combinative capabilities, by combining different pieces of existing knowledge (Kogut and Zander,

1996; Kogut and Zander, 1992), which presents a powerful motive for seeking alliances for innovation. Alliances could be construed as efficient vehicles for knowledge transfer across organizational boundaries – knowledge that is difficult to grasp because of its tacitness (Kogut, 1988; Inkpen, 1999; Teece, 2000). However, transactional concerns are quite intense in alliances (Mowery *et al.*, 1996; Inkpen and Beamish, 1997) and this could potentially make learning in an alliance a competitive process, as partners race to learn each other's domain of knowledge (Hamel, 1991; Anand and Khanna, 2000; Khanna, Gulati and Nohria, 1998).

Several researchers have focused on the factors that increase learning in alliances. Learning in IJVs requires both diversity and absorptive capacity (Lyles, 1988; Lyles and Salk, 1996; Parkhe, 1991; Inkpen, 1999; Dussauge, Garrette and Mitchell, 2000). While diversity suggests unrelated knowledge bases, absorptive capacity would point to related knowledge bases among the partners. Here is the learning paradox of alliances where the most efficient alliance combination becomes the least effective learning. Research has also highlighted the need for shared control in IJVs to facilitate learning and performance (Lyles and Salk, 1996; Dhanaraj and Beamish, 2002). As one would expect, learning is also influenced by the knowledge characteristics such as complexity, tacitness, and ambiguity (Simonin, 1999; Badaracco, 1991; Anand and Khanna, 2000).

A more recent development in learning research focuses on the role of social capital in learning. Kogut and Zander (1996) suggest that firm boundaries in essence provide a context for knowledge to be developed and exploited, and emphasize the role of coordination, identity and learning as key activities of the firm. In a similar vein, Nonaka (1994) emphasizes the role of 'socialization' in the diffusion of knowledge within organizations. Van den Bosch *et al.* (2001) suggest that socialization capabilities, by which is meant the ability of the firm to produce a shared ideology that offers members an attractive identity as well as collective interpretations of reality, may influence learning by specifying broad, tacitly understood rules for appropriate action under unspecified contingencies. Kale *et al.* (2000) suggest that relational capital between partners enhances learning capability, simultaneously mitigating the transactional concerns. Parkhe (1998) and Inkpen and Dinur (1998) suggest the role of trust in learning between partners in an alliance. Lyles and Salk (1996) suggest that informal communication is critical for knowledge acquisition. Makhija and Ganesh (1997) argue that through social networking processes, the highest order learning and the most tacit knowledge becomes shared. All these conceptual ideas refer to the central role of the social or community orientation of knowledge, an aspect that Cohen and Levinthal's (1990) conceptualization of absorptive capacity does not explicitly mention. Recent work on social capital (Cohen and Prusak, 2001: 4) provides a comprehensive conceptual framework:

> Social capital consists of the stock of active connections among people, the trust, mutual understanding, and shared values and behaviors that bind the members of human networks and communities and make cooperative action possible.

They suggest trust, loyalty and membership as three dimensions of social capital. This provides a parsimonious and efficient operationalization of the social aspects of learning. However, the cognitive aspects such as absorptive capacity and its relationship to social capital and the learning processes need to be explored further. Brown and Duguid's (2001) observation that knowledge flowing across a group is a function of the membership of the group, whether they are bound by practice or by organizational bonds, fits well with the theory of social capital, and thus may provide an anchor to develop a unified theory of organizational learning.

Organizational learning: a process model

One assumption about organizational learning is that knowledge can be transferred between individuals, teams, and organizational units. To understand the dynamics of learning, Figure 5.1 presents two levels of learning, lower and upper, which are closely linked by processes and histories within an organization (Lyles, 1988). Lower-level learning, or explicit knowledge, is the result of repetition and routines. It is the knowledge that can be explained and codified. It results in standard operating procedures or success programmes or in new management systems that handle repetitive, unchanging situations. Higher-level learning involves an adjustment of overall missions, beliefs and norms, resulting in new frames of references, new skills, and an unlearning of past success programmes. It is the tacit, sometimes unconscious knowledge that relies on the organizational memory and discrimination skills of the organization. The organizational knowledge structures include the storage of the belief systems, memories of past events, stories, frames of reference, or values. For example, the organizational knowledge structure may include a story about the company president and

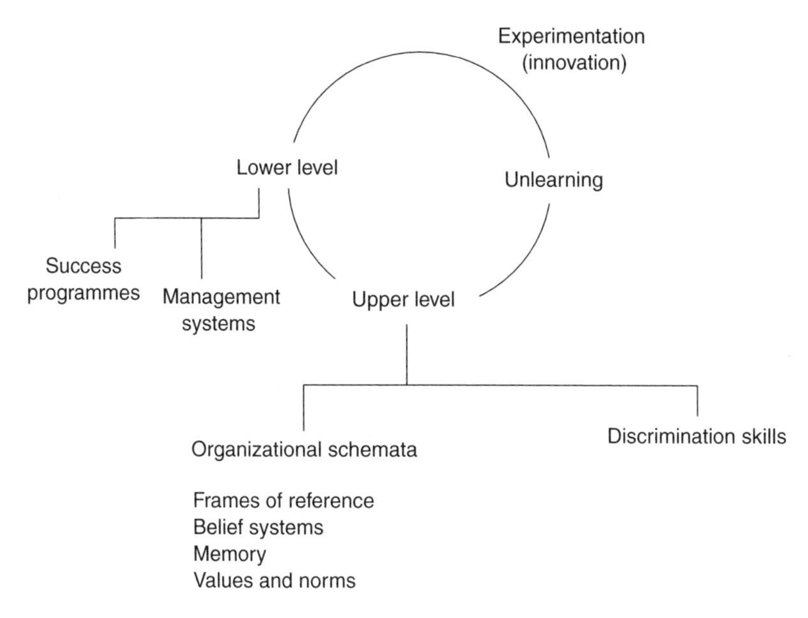

Figure 5.1 Learning framework (Lyles, 1988: 304)

how the president handled a very tough situation that is repeatedly told around the company – it eventually serves as a guide for future behaviour (Levitt and March, 1988). Scholars have shown that organizational stories serve as important 'lessons' that are stored in organizational memory. Since upper-level learning is to a large extent tacit knowledge, it is impossible to articulate all know-how included in upper-level learning, and it is practical intelligence (Polanyi, 1966).

A unified framework for organizational learning

The wide array of research suggests an increasing interest in pursuing research on organizational learning and international alliances. Unfortunately, research in this area does not seem to converge but there has been proliferation of constructs and models. The field is still dichotomized in its focus: structure *vs* process orientation, cognitive *vs* social aspects of learning, and knowledge characteristics *vs* organizational characteristics. While it is very fashionable to talk about knowledge management or organizational learning, very little has been done to develop research that can significantly influence managerial practice.

Given the phenomenal contributions over the last decade, both theoretical and empirical, we suggest that it may be worthwhile providing a unified framework to facilitate cumulative knowledge development and advancing the field.

As discussed earlier, Lyles' (1988) process model provides a starting point for this, with the emphasis on the learning processes and capabilities. We integrate this with (a) knowledge-based theory of the firm (Nelson and Winter, 1982; Winter, 1987; Grant, 1996); (b) theory of absorptive capacity (Cohen and Levinthal, 1990); and (c) theory of social capital (Cohen and Prusak, 2001; Cohen and Levinthal, 1990). In developing this unified framework, we postulate four fundamental principles:

1. Organization learning is an outcome of six distinct processes, namely, identifying, discriminating, experimenting, reflecting, unlearning and communicating – which also form key learning capabilities for an organization.
2. Absorptive capacity of an organization sets the cognitive context for learning, which constrains and guides the learning processes, and dictates the technological domain where the learning capabilities could potentially be deployed. Absorptive capacity is a cumulative stock of prior learning within the organization.
3. Social capital of an organization sets the social context for learning that can accelerate or decelerate the learning processes, and dictates the organizational domain where the learning capabilities could potentially be deployed. Social capital is a function of the cumulative stock of trust, loyalty and memberships that exist in the organization.
4. Organizational factors (knowledge activist, organizational structure, and reward systems) can serve as a catalyst to the learning process (Lyles and Salk, 1998; Lane and Lubatkin, 1998; Van den Bosch *et al.*, 2001), influencing or impeding the learning capabilities of the organization.

Figure 5.2 presents a schematic flow of our unified theory of organizational learning. In the following sections we develop the model and highlight the critical implications.

We build on the process model of organizational learning (Lyles, 1988). We suggest that an organization learns through a series of six processes, which may be present in varying degrees. The more tacit the knowledge is, the more pervasive these steps are. Although we present it as a sequence of steps, the process can begin anywhere. Learning can begin as an organization identifies a problem or an issue to be

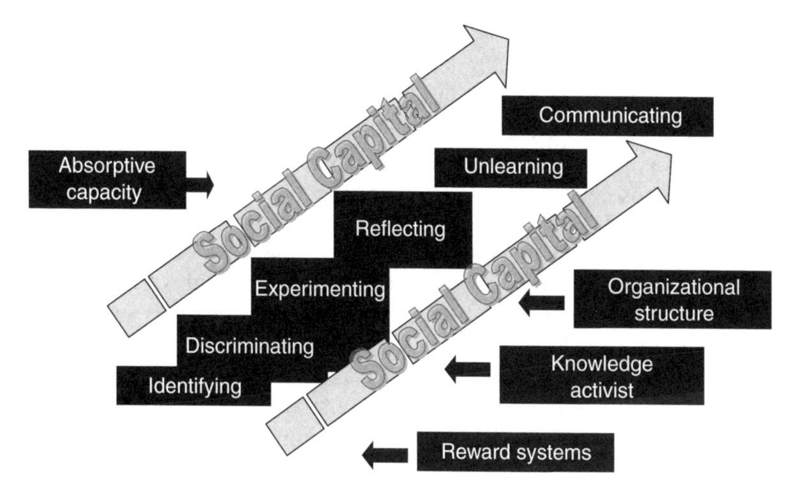

Figure 5.2 A structure–process view of organizational learning

resolved – which seems to be a common starting point, or can begin as an organization reflects on a recent disastrous performance or a failed sale.

Identification

While accidental learning occurs frequently within an organization, purposive learning is the norm in many organizations. Identifying what needs to be learned is a task in itself. Organizations that do not know what they do not know may not be in a position to identify new areas that need to be learned. In many international alliances, learning objectives evolve over the negotiation period. In alliances where one partner contributes technological know-how and the other partner provides the market linkages, learning objectives are implicit. Unless the IJV makes such learning objectives explicit for its managers, it may not be able to achieve its objectives, and quite often when the desired learning does not take place in an alliance, it shows poor performance. Hamel (1991: 90) notes that learning intent with which partners move into an alliance is a key determinant of learning:

> Insofar as it could be ascertained, the Japanese counterparts in these alliances seemed to possess explicit learning intents – with one possible exception. This apparent asymmetry in collaborative goals between Western and Japanese partners is deemed significant because in no case did systematic learning take place in the absence

> of a clearly communicated internalization intent. In cases where one partner had systematically learned from the other, great efforts had been made to embed a learning intent within operating level employees.... In one firm where learning did not take place the blame was put on a failure to clearly communicate learning objectives to those with inter-organizational roles.

Unfortunately, very little systematic research has been done in this area as to how organizations discover issues to learn. Nonaka and Takeuchi (1995: 74) note that organizational intent or its aspiration is a condition for knowledge creation:

> The most critical element of corporate strategy is to conceptualize a vision about what kind of knowledge should be developed and to operationalize it into a management system for implementation.

Urban and von Hippel's (1988) work on lead customers as sources of innovation suggests that relationships and communication are key aspects of identification for learning.

Experimentation

Learning suggests 'newness' and 'unfamiliar' knowledge. Acquiring such knowledge often demands that organizations or employees go outside their familiar zones and try out something new. Experimentation involves a systematic searching for and testing of new knowledge. There is always a 'first' time when an idea is applied within the organization and depending on where and when the idea had been tried outside the organization, there is an element of uncertainty. Often when new ideas come from external sources, the 'Not-Invented-Here' syndrome can be a major issue to overcome before it makes it to practice. Fear of failure and many times pure callousness can kill ideas. An organization 'learns' when it 'experiments' with unfamiliar ideas. In many cases, it could be a 'guided experiment' wherein the organization that learns is instructed by an individual or team external to the organization. For example, in many international alliances, a foreign partner typically contributes to the technological know-how, and the joint venture organization acquires the technical know-how. Even in such a case, it is an experiment for the JV as it acquires the know-how by doing. As the tacitness, complexity and ambiguity of the knowledge increases, experimentation will become a critical component of the learning process (Arrow, 1962). It is here that many innovations, both new practices and products, find their origin. However, experimentation, to be fair, would also lead to several unsatisfactory outcomes, which may sometimes be characterized as 'failures'. However, these become valuable as they guide the subsequent search process for a learning organization. In new created joint ventures, there is a large scope for experimentation, as typically several systems – production or marketing systems – can be implemented in the joint venture. To managers drawn from one parent, the systems from the other parent are 'new' and implementing them requires experimentation. One way to illustrate the role of experimentation in learning is to consider how one learns driving. Whether you have a formal tutor or not, learning to make a right-angle turn is a matter of several experiments. There may be some who might get that skill within one or two turns, but for some others it may require several attempts before they are able to make that turn. No amount of instruction can help until the learner takes that step of turning. Even wrong outcomes from an experiment are helpful because they provide a feedback for the subsequent experiment. An organization that is adept at experimentation would have an incentive system that favours risk-taking. As Garvin (1993) observed:

> Employees must feel that the benefits of experimentation exceed the costs; otherwise, they will not participate. This creates a difficult challenge for managers, who are trapped between two perilous extremes. They must maintain accountability and control over experiments without stifling creativity by unduly penalizing employees for failures.

It is this process of experimentation that links learning and innovation. Experimentation could be viewed at two different levels. First, experimenting with an idea that has never been put to test in any known context, where the uncertainty can be high. Second, experimenting with an idea that has been tested under different context, where the uncertainty may be zero or low. In both these cases, the fundamental processes of experimentation are quite similar. Innovation is a unique outcome of learning, as organizations become adept at experimentation. Discovery of new problems and new solutions happen as organizations take previously unexplored paths in learning. Failure or discovering an unworkable idea should be an equally acceptable outcome in experimentation. It is hard to harness innovation when 'failure' is not allowed. Edison, after conducting one thousand experiments in search of a filament material for his lamp proudly declared that he then knew 1000 ideas that would not work, which made it easier for him to find that next one that would work. Systematic research in the area of new product development area suggests that the knowledge gained from failures is often instrumental in achieving subsequent successes (Maidique and Zirger, 1985).

Reflection

Organizations lose many opportunities for learning because they do not reflect on the past. George Santayana's famous phrase 'Those who cannot remember the past are condemned to fulfil it', seems to be a fact of life in many organizations. Both intentional experiments and unintended organizational actions provide rich contexts for reflection. Organizational reflection is an honest introspection of the past behaviour and performance (Brookfield, 1995; Dewey, 1933; Mezirow, 1991). It is a process whereby we carefully consider the knowledge, beliefs, assumptions, actions and processes that influence our behaviour in order to understand our experiences. It involves reflecting on the content, process or premise of an underlying issue (Mezirow, 1991). Content reflection focuses on the content or description of a problem or issue, process reflection analyses the methods and strategies that are being used to resolve the problem, and premise reflection focuses on the underlying assumptions or beliefs and consideration why the problem is a problem in the first place. Often periodic reviews and performance assessment meetings become a context for 'fixing the blame' rather than seeing it for what is – a rich context for learning. Schon (1983) suggests that managers and organizations can learn a lot as they do, by taking time to reflect on what they have done. Organizational reflection is the process of taking time to review the successes and failures, assessing them systematically, and recording the lessons in a form that employees find open and accessible (Garvin, 1993). In order to be effective, organizations must be willing to reflect on both successes and failures. In organizations where honest discussion of mistakes is not encouraged, there would be a tendency to hide the mistake, and the factors that led to it and to the outcome never coming to the surface, guaranteeing the possibility of reoccurrence of the same mistake in another place. Shaw and Perkins (1991) note that performance pressure, which makes time for reflection a luxury and ill-afforded, or the competency trap, which suggests it is quicker and easier to keep doing what is already being done even if it is not in the best interests of the organization, and absence of learning forums or structures indicate a leadership and culture that does not reward learning.

Reflection often allows new perspectives to enter the organization based on experiences. One characteristic of knowledge is that it requires time to be assimilated, typically referred to as 'soaking time'. In addition, when the action is thought through outside the context, in a systematic way, it unearths several fundamental issues – including mental models and critical assumptions (Preskill and Torres, 1999). Organizational reflection enables an organization to re-enact the past, and open up the behaviour for challenge either by peers, or superiors, or by outside consultants (Schon, 1983). It enables the organization to interpret individual behaviour within a holistic framework (Preskill and Torres, 1999). Individual experiences are communicated through a process of dialogue, allowing practitioners to confront contradictions which otherwise might go unchallenged and unquestioned, and helping uncover assumptions and beliefs. Senge (1990: 353) defines dialogue as

> A sustained collective enquiry into everyday experience and what we take for granted. The goal of dialogue is to open new ground by establishing a container or field for enquiry, a setting where people more aware of the context around their experience and of the processes of thought and feeling that created that experience.

It is through a process of dialogue that reflection, from being an individualistic process, becomes an organizational capability.

Discrimination

This was discussed in our earlier model as a process of identifying directions to go based on the experimentation and reflection. Discrimination is the art of assessing the 'existence' of knowledge, its 'newness' and its 'relevance' (Lyles, 1994). Discrimination skills are another form of upper-level learning, involving the ability to discern differences among situations and to choose the appropriate course of action (Lyles, 1994). Levitt and March (1988) identify the importance of discrimination skills through simultaneous evaluation of routines that lead to successful outcomes. By discrimination, we mean the ability to segment action into situations where one series of actions would be appropriate versus another set of actions. It involves the simultaneous evaluation of alternative courses of action to determine if one is more appropriate under a set of circumstances than another. It implies that organizations can distinguish between situations and make choices about appropriate actions (Lyles, 1994: 24):

> A simple example might be, how does one discriminate between apples and oranges? There are two separate and distinct behaviors for eating each. Both are fruit. They are both round. One learns to discriminate on the differences between them. The colors are different, and the textures are different. Over time one learns to discriminate between apples and oranges and to learn the appropriate 'success program' for each.

Discrimination is affected by how past experiences have been encoded into the organizational memory, past success programmes, the amount of time available, the number of people involved, situation assessment and salience (Lyles, 1994). While commitment to past success programmes and the number of people involved negatively influence discrimination capability, organizational memory, time and salience influence it positively. One way discrimination skills are used is in determining how to enter a foreign market, either through a JV, a wholly owned subsidiary or a distributor. A company's knowledge of the market and its own strengths is needed to make the decision and it will change from country to country and from situation to situation. This involves adaptation of previous success programmes to the current situation. Limited research has been done in this area and advancing the field of organizational learning would require explicit operationalization of this capability and linking it to the other learning processes.

Unlearning

Unlearning involves the process of reframing past success programmes to fit with changing environmental and situational conditions. It is triggered by mistakes, failures, organizational change or poor performance. It involves taking solutions espoused by experts and making adaptations. The organizational unlearning construct has been widely recognized as a fundamental component of organizational learning in the practitioners' literature and in many keystone articles on learning in the academic literature (Argyris and Schon, 1974; Hedberg, 1981; Nystrom and Starbuck, 1984; Bettis and Prahalad, 1995; Lengnick-Hall and Wolff, 1999). Yet, despite this recognition, theory development around the unlearning construct has languished and no substantial empirical work has ever been done on strategic unlearning, prompting authors to call for more substantial work on the subject (Huber, 1991; Bettis and Prahalad, 1995). Hedberg (1981) defines unlearning as 'a process through which learners discard knowledge. Unlearning makes way for new responses and mental maps'. Unlearning can be intentional, as a result of concerted organizational change efforts, or unintentional, as a result of employee turnover or organizational 'forgetting' (Argote, 1996). Bettis and Prahalad (1995: 10) provide a basis for discussion by stating that learning is a function of previous unlearning in the equation:

$$L_t = f[F_{(t-1)}]$$

where L_t = learning in period t, $F_{(t-1)}$ = unlearning in period $t-1$, and 't' can be thought of as small. Although, this equation seems intuitive, discussion in the literature suggests that 't' cannot be automatically thought of as small (Postman and Stark, 1965). Bettis and Prahalad (1995) acknowledge that the temporal relationship between learning and unlearning is irregular, and Hedberg (1981) points out that organizations can become paralysed by the uncertainty and confusion that results from the failure to learn new practices after old ones have been discarded. Argyris and Schön (1974) describe the unlearning process in individual managers as though it was an inherent part of the learning process. Specifically, they discuss the defence mechanisms that protect established beliefs and knowledge, and prevent learning, called 'defensive reasoning'. Individuals use defensive reasoning to protect their senses of superiority and being in control, and to suppress negative feelings associated with failure and job insecurity. The defensive reasoning process is described as a self-perpetuating, closed cycle, which Argyris calls the 'doom loop'. He suggests that the doom loop can be broken through careful and honest reflection, and through social interaction in which questioning others ideas is culturally acceptable. Thus, unlearning in organizations is an inherently political process in which power and influence are asymmetrically distributed across organizational members. The knowledge and behaviours to be unlearned by the organization may play a significant role in how the unlearning/learning process unfolds. Hedberg (1981) identifies three targets of unlearning in individuals: (1) mechanisms for identifying and selecting stimuli; (2) connections between stimuli and responses; and (3) connections between multiple responses that make up complex behaviour. The problem of unlearning is not only a cognitive problem, altering perceptual maps, but a problem of driving out old behaviour with new behaviour. Unlearning can become a significant hurdle for a laggard attempting to compensate for past skill failure (Hamel, 1991).

Communication

Communication in the process of learning can be viewed as a sense-making process to create knowledge as well as an information distribution process to share knowledge (Daft and Weick, 1984). This is a process of sharing among the members of the organization, problems, search processes, and solutions – both successes and failures. Organizations often do not know what they know (Huber, 1991). Communicating is a way of interpreting information, as Daft and Weick (1984: 294, 296) suggest, 'the process through which information is given meaning' and also as 'the process of translating events and developing shared understandings and conceptual schemes'. Communicating assumes existence of some common schemata and patterns under which scientific knowledge can be transmitted. Brown and Duguid's (1991, 2001) work on communities of practice suggests that communication among the members of the organization provide a way of resolving problems and creating new knowledge. The common schemata and language that members of similar practice develop within their group helps to communicate ideas that can not be easily articulated. The informal socialization that exists among the group helps to specify broad, tacitly understood rules for appropriate action under unspecified contingencies (van den Bosch *et al.*, 2001).

Communicating across or within organizations assumes receptivity, i.e., willingness to access, receive and process information that is shared. This becomes complex when groups from different parts of the same organization or different organizations need to communicate with one another (Szulanski, 1996). Hamel (1991), in his research on learning in international alliances notes that it is possible that the partners in an alliance could have differing levels of receptivity:

> When we saw [our larger Western partner] doing something better, we always wanted to know why. But when they come to look at what we are doing, they say, 'Oh, you can do that because you are Japanese,' or they find some other reason. They make an explanation so they don't have to understand what we are doing differently.

Hamel notes that an abundance of resources and a legacy of industry leadership, whether real or perceived, makes it difficult for a firm to admit to it that it had something to learn from a smaller partner. Communication also necessitates transparency. He also notes that while intent establishes the desire to learn, transparency determines the potential for learning. Transparency allows organizational members to share freely.

Structural factors of learning capability

The structural factors that enhance learning capabilities can be broadly classified into three broad categories: (1) Absorptive capacity (cognitive domain); (2) Social capital (social domain); (3) Knowledge structure (technical domain); (4) Organizational systems (people domain). In this section, we highlight these four structural factors, and provide an overview of their role in learning.

Absorptive capacity and social capital

Absorptive capacity, widely discussed in the literature, has been the central construct in organizational learning literature (Cohen and Levinthal, 1990; van den Bosch *et al.*, 2001). Absorptive capacity is a measure of the prior knowledge accumulated in the organization. Theorists suggest that it is difficult for organizations to identify, understand, and assimilate new knowledge without some prior related knowledge. While it is true that it is impossible to absorb new unrelated knowledge without a minimal stock of prior knowledge, it is also true that the presence of the requisite level of prior knowledge does not guarantee learning, as other social factors may inhibit the process. We suggest that absorptive capacity is a necessary but not sufficient condition for learning to occur.

If we accept the notion that 'learning is both action outcomes and changes in the state of knowledge' (Lyles, 1988: 302) then it becomes not only a cognitive process but a social process too. Cognitive domain may lie dormant and useless until a matching social domain is found. Relationship among organizational members, as well as their relationship with customers and vendors dictate what is being learned and how well it is being learned (von Hippel, 1994; Brown and Duguid, 2001; Argyris and Schon, 1974; Hamel, 1991; Cohen and Prusak, 2001). The role of social capital in organizational learning is particularly conspicuous in alliances. Gulati (1995) suggests that trust enables firms to reduce dependence on equity structures to govern the relationship and Zaheer *et al.* (1998) suggest trust reduces negotiating costs in alliances and also enhances alliance performance. Ring and Van de Ven (1992) suggest that personal connections and relationships between contracting firms play an important role and suggest the term 'relational capital' among the partners. The strength of the relationship between alliance partners provides an organizational climate, where there are common linguistic schemata for communicating tacit knowledge, reciprocity, receptivity and transparency between the partners. These are essentially the same issues when one looks at learning within an organization, as discussed in the earlier section here. In attempting to develop a unified framework, we follow Cohen and Prusak (2001) to use the concept of social capital, as the social domain of learning. Social capital enhances the learning capabilities of the firm. In an organization where there is a high social capital, there will be increased levels of learning processes, and vice versa.

Organizational systems – people domain

While absorptive capacity and social capital refer to the cognitive and social domains of an organization, and point to its learning capabilities and its ability to influence the learning processes in an organization, research suggests that an organization can influence these by using three organizational means:

- knowledge activist (Lyles, 1988);
- organizational structure (Lane and Lubatkin, 1998); and
- reward systems (Lane and Lubatkin, 1998).

The knowledge activist performs the role of a champion and often the initiator and a catalyst of the learning processes (Howell and Higgins, 1990). Organizational structure influences how a firm processes knowledge (Kogut and Zander, 1992). Van den Bosch *et al.* (2001) discuss the role of different types of organizational structures on the learning capability of the organization. Finally, the reward system can be a motivating factor for accelerating learning in organizations (Huber, 1991).

An application of the learning framework to two IJV cases

In this section, we want to analyse two previously published cases drawn from previously published material from other researchers that illustrate our organizational learning model: Wil-Mor Technologies (Inkpen, 1999) and Toppan Moore Co. (Beamish and Makino, 1992). These case studies show two contrasting organizational learning behaviours in international joint ventures (both these cases have one Japanese partner) and the factors that influence learning.

Wil-Mor Technologies, Inc.

This case describes a joint venture between Wilson, an American manufacturer of plastic and metal parts for the automotive and appliance industries, based in Detroit, and Morota, a Japanese manufacturer of automotive parts. Both these firms were of similar size recording sales of about half a billion US dollars in the mid 1990s. While Wilson's clientele was American auto-makers, Morota's clientele was Japanese auto-makers, with Toyota forming the largest component. The rapid growth of the Japanese auto-makers in the US market provided a need for Wilson to find an opportunity to expand their customer base, as well as an opportunity for Morota to expand into the US. Morota's need for an American partner coupled with its access to Toyota and other Japanese auto-makers matched quite well with Wilson's American presence coupled with its need for developing new customer base. After six months of negotiation, both the firms signed a 50–50 joint venture in early 1993, with Wilson taking over the responsibility of the management of the workforce and Morota the equipment installation, engineering support, worker training and liaison with Toyota to ensure sales orders. Wilson saw the JV as a means of expanding its market share as well as an access to a growing segment of the market, along with allowing it to learn from its Japanese partner. The management team was contributed by both partners. The JV president was nominated by Morota, along with an engineering manager, quality manager, and a marketing manager. Wilson contributed the general manager (who reported to the president) and an operations manager, human resource manager, controller and a marketing manager:

> The JV did not begin smoothly. The Japanese managers insisted on complete technical responsibility. Johnson, the general manager, was not allowed to assist in the technical setup. This caused several problems because the Japanese were unfamiliar with many of the basic aspects of establishing a new plant, especially one in North America. The Japanese insisted on running the operation their way. They used a Japanese approach in selecting suppliers. Johnson estimated he could have saved the JV about $300,000 a year if a North American approach had been used to select suppliers. However, the Japanese insisted that, if possible, suppliers should be selected not just on the basis of price but because they had established themselves as capable suppliers to Wilson or Morota. (Inkpen, 1999: 418)
>
> Johnson (general manager, appointed by Wilson) became convinced that the Japanese managers were deliberately excluding him from the management process. The Japanese managers would regularly hold meetings and exclude the Americans. When meetings were held with both Americans and Japanese present, they would last for hours because of the necessity to translate from English to Japanese. In addition, the Japanese managers corresponded daily by fax with their head office in Japan and would meet socially in the evenings

> and on weekends. The inevitable result was two distinct management 'camps': the Americans and the Japanese.... Very few decisions in the JV were 'joint' because the Americans and Japanese rarely talked to each other. (Inkpen, 1999: 410)

The management team was changed in response to the internal situation and the new team (Kawajima, as president, from Morota and Easton, as general manager, from Wilson) both had international experience, and were fluent in English:

> The new management team got off to a much better start than the previous one. Both Easton and Kawajima were avid golfers. They began playing golf together regularly and involved several of the other managers. Gradually, the tension between the American and Japanese 'camps' began to ease.... By early 1996, Wil-Mor had successfully bid on several General Motors contracts.

Despite the renewal in the joint venture, the realization set in that in order to reach the desired profitability levels it would take more time and more investment in expansion. While Wilson was worried about the ongoing operating loss, Morota did not perceive it as a serious issue and was content on its accomplishments in quality and growth. The JV ended in 1997 with Wilson reducing its stake to 5 per cent and selling the remainder to Morota. Unfortunately, Wilson never used the opportunity to absorb valuable process capabilities from Morota, which would have helped to increase its overall product quality.

Toppan Moore

This case describes a joint venture started in 1965, between Toppan Printing, a Japanese printing company and Moore Corporation, a Canadian manufacturer of business forms. Toppan operated nine plants in Japan and had subsidiaries in Australia, Hong Kong, Indonesia, Korea, Singapore and the United States. Toppan's desire to move in to the business forms industry and its realization that the business forms production and marketing were quite distinctly different from its capabilities motivated it to seek a partner in Moore. Moore also recognized that customer behaviour in Japan might be different from what it was used to in North America and that Toppan being a leader in Japan, could be a significant factor in success in that market. Toppan Moore was set up as a 55 : 45 joint venture with Toppan taking the majority ownership owing to a government restriction on foreign ownership. Toppan Moore's management was entirely drawn from Toppan, with the exception of a vice-president, appointed by Moore, who served as a 'communications pipeline' between the two parent firms. Moore was generous in providing support to the local operations. Yamada, Toppan Moore's president:

> Moore sent us their newest machine, a high performance press that cost almost ¥75 million, more than the cash assets (¥70 million) of the whole company. Even though Moore knew that we couldn't afford this, they didn't send a cheaper, lower performance model. Our general managers were very impressed that Moore sent us their best equipment when they didn't have to, and they didn't expect quick payment. In the end, we did somehow raise enough money to pay for it... Moore was generous with their technology. Over the years, they made a great contribution to Toppan Moore's production technology and production management skill. The company showed very human feelings.

The joint venture agreement contemplated on joint efforts on R&D, product design, quality and manufacturing, and there was frequent communication at every level between the companies on a whole range of issues, including exchange of engineers between the two companies for short periods. Shortly after the JV was established several of its managers visited two of Moore's US plants to assess their sales and production systems. Moore's sales director conducted a training seminar at Toppan Moore for the sales force and managers:

> Moore had particular ideas that they wanted to bring into the joint venture, regarding both sales and production methods. Moore wanted the sales force to work under a territory coverage system based on commissions. As well, Moore wanted to introduce a 3-shift system and to control production using cost-based pricing. There was some minor resistance from the plant workers when we added a third shift, but this was largely because of a reduction in net wages.... We did have some early disagreements over pricing, however. Moore used a highly disciplined pricing scheme based on formal planning and cost benefit analysis. In Japan, many companies put priority on expanding market share, and prices tended to fluctuate. Under Moore's system, we did much less price-cutting. This is just one aspect where we adapted our operating methods. We were never forced. As the company developed, we were able to select which methods we wanted to incorporate into our own practices.

From 1965 to 1990, the joint venture had grown from ¥800 million to ¥148,500 million and the employees grew from 410 to 2774. Several new business operations emerged from the JV, such as Toppan Moore Operations, founded in 1975 to provide technical services to client companies, and Toppan Moore Learning, a JV between Toppan Moore and another US-based company launched in 1980 to provide media products for educational institutions. Mr Ogura, Toppan Moore's managing director:

> So far, Toppan Moore has enjoyed immense success. This is not the norm for many joint ventures in Japan. One of the reasons is that Moore provided good circumstances for the development of the company. ... They gave us a lot of autonomy ... We make decisions on personnel, investment, and fund raising without detailed consultation. We are able to manage freely, and we have adopted many Japanese principles, such as a long-term focus, interdependence among companies, business diversification, and a management style based on loyalty and human feeling. Toppan Moore is very much a traditional Japanese company.

Discussion

The cases are provided here only to illustrate the framework, not to support the framework. Since these are post-hoc interpretation of events, we do not claim empirical support with these cases. Nevertheless, these cases do provide how the cognitive, social and organizational domains of organizational learning interact with each other and influence the learning processes in an organization. Table 5.1 summarizes the case data against our theoretical framework. In the Wil-Mor case, one observes a low level of learning process, coupled with low social capital, and organizational systems. The presence of high levels of absorptive capacity has not been of much help in promoting learning. In the Toppan Moore case, one sees how the various learning processes come together at work, and how social capital enables these processes to go on. One can also see the effective role of the organizational systems in promoting learning.

We have raised three connected issues that relate to the theory of organizational learning and its application to international alliances. First, we question the theory of absorptive capacity as the central pillar for organizational learning theory. We suggest that absorptive capacity provides the base condition for organizational learning to take place, and it also constrains what can be learned and in what pace it can be learned. It is a necessary condition for learning for not a sufficient condition.

Table 5.1 A comparison of Wil-Mor Technologies and Toppan Moore

	Wil-Mor Technologies	*Toppan Moore*
Learning processes		
Identification	Low. No attempt was made to explicitly identify the need for learning the process technology or the market linkage.	High. Moore identifies what need to be transferred to Toppan, and Toppan Moore also takes the initiative to provide the 'wish-list'.
Experimentation	Low. Opportunities for experimentation were not made use of; nor were new opportunities sought (e.g., Wilson had a number of other plants within the US, where transferring the process technology would have yielded a sizable return).	High. Experimenting with the commission-based and territorial sales system was a risky one for the IJV, given the culturally different environment.
Discrimination	No clear evidence.	High. The IJV shows an active role in isolating the relevant knowledge systems from Moore, as well as from Toppan. The JV keeps the sales and marketing system of Moore and the personnel system of Toppan.
Reflection	Low. The deteriorating organizational climate was left unattended for the first three years.	High. Quarterly reviews and personnel visits, and management reviews.
Unlearning	No clear evidence.	High. The IJV gives up the Japanese sales system to adopt the new one from Moore.

Table 5.1 (Continued)

	Wil-Mor Technologies	*Toppan Moore*
Communication	Low.	High. Both within the IJV and with the parent (Moore) who is the source of the technology.
Social capital	Low in the first three years (two separate organizations despite being an IJV). Turnaround in the latter years with the new management – the joint golfing programme, etc.	High. High level of trust, loyalty and membership – consistently building it over years as the operations expand.
Absorptive capacity	High. Similar industry.	High. Similar industry.
Organizational structure	Hierarchical.	Networked – particularly the role of the board of the Toppan Moore – and highly connected.
Knowledge activist	Low – for the first three years. High – with the new GM.	The GM of the IJV becomes the chief knowledge activist.
Reward system	No direct incentive for learning.	Not clear.
Overall learning	Very low.	Very high.

Absorptive capacity contributes to learning by influencing the processes of identification, discrimination, experimentation, reflection, unlearning and communication. The impact of absorptive capacity on learning effectiveness can only be understood by considering the processes through which it works. We have highlighted here the social theory of learning, which emphasizes the role of social capital (trust, loyalty and membership) as a critical element for any of the learning processes to take place. Thus, a unified framework that integrates the cognitive and social process views of learning would enhance our understanding of organizational learning and help to create new knowledge to enable businesses to remain competitive in a rapidly changing market.

Second, we have highlighted here the close link between the two research streams: alliances and organizational learning. Irrespective of the type of the alliance, an alliance invariably involves sharing of competences across organizational borders and learning is paramount to the success of alliance. An international alliance, at best can present a classic case of discontinuous learning, where an organizational has to learn something that is socially disconnected to it. In essence, it is a special case of organizational learning. A unified theory of organizational learning, which can be applied to both singular organizations and

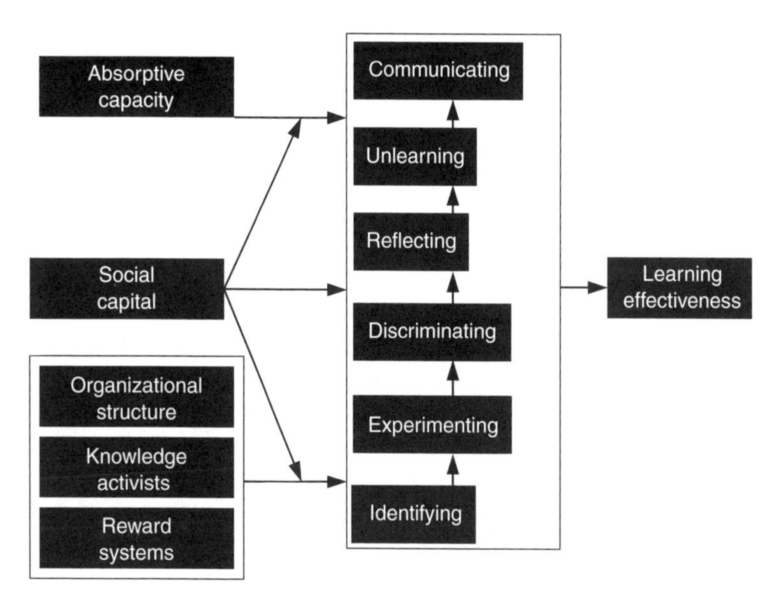

Figure 5.3 An integrated process–structural model of organizational learning

alliance organizations, may provide us a better direction to pursue for further research as well as practice.

Third, we have presented a case for integration of the structural and process elements of organizational learning, which presents a more realistic picture of the actual process. While process elements focus on a series of processes that eventually become capabilities ingrained in the organization, structural elements such as absorptive capacity and social capital work to enhance the effectiveness of these learning capabilities. This approach would provide a better understanding of the social and cognitive factors, as well as the structural and process factors of organizational learning. Empirical studies need to be done to validate such a conjecture. With the growing number of international alliances, this would be an opportune study to pursue. Figure 5.3 presents a possible model for an empirical study.

Conclusion

We have presented a unified theory of organization learning, drawing on the process view of learning, knowledge-based theory of the firm, absorptive capacity and social capital. The framework provides an opportunity for researchers to integrate cognitive and social domains of learning. This will be helpful to develop some practical insights to enable organizations to become competitive in the twenty-first century.

Note

1. This material is based upon work supported by the US National Science Foundation under Grant No. 0080152. Any opinions, findings, and conclusions or recommendations expressed in this material are those of the authors and do not necessarily reflect the views of the National Science Foundation.

References

Anand, B. N. and Khanna, T. (2000) Do firms learn to create value? The case of alliances. *Strategic Management Journal*, 21(special issue): 295–315.

Argote, L. (1999) *Organizational Learning: Creating, Retaining, and Transferring Knowledge*. Boston: Kluwer Academic.

Argyris, C. and Schön, D. C. (1974) *Theory in Practice: Increasing Professional Effectiveness*. San Francisco, CA: Jossey-Bass.

Arrow, K. J. (1962) The economic implications of learning by doing. *Review Econometric Studies*, 28: 155–73.

Badaracco, J. L., Jr (1991) *The Knowledge Link*. Boston: Harvard Business School Press.

Beamish, P. W. and Makino, S. (1992) *Toppan Moore*. Teaching Case and Teaching Note. London, Canada: Ivey Management Services, University of Western Ontario.
Bettis, R. A. and Prahalad, C. K. (1995) The dominant logic: retrospective and extension. *Strategic Management Journal*, 16: 5–14.
Bosch, F. van den, Volberda, H. W. and Boer, M. de (2001) Coevolution of firm absorptive capacity and knowledge environment: organizational forms and combinative capabilities. *Organization Science*, 10: 551–68.
Brookfield, S. D. (1995) *Becoming a Critically Reflective Teacher*. San Francisco, CA: Jossey-Bass.
Brown, J. S. and Duguid, P. (1991) Organizational learning and communities of practice: toward a unified view of working, learning and innovation. *Organization Science*, 2(1): 40–57.
Brown, J. S. and Duguid, P. (2001) Knowledge and organization: a social-practice perspective. *Organization Science*, 12(2): 198–213.
Cohen, D. and Prusak, L. (2001) *In Good Company: How Social Capital Makes Organizations Work*. Boston, MA: HBS Press.
Cohen, W. M. and Levinthal, D. A. (1990) Absorptive capacity: a new perspective on learning and innovation. *Administrative Science Quarterly*, 35: 128–52.
Daft, R. L. and Weick, K. E. (1984) Toward a model of organizations as interpretation systems. *Academy of Management Review*, 9(2): 284–95.
Dewey, J. (1933) *How We Think*. New York: Heath.
Dhanaraj, C. and Beamish, P. W. (2002) Ownership strategies and survival of Japanese overseas subsidiaries. *Best Paper Proceedings of the Association of Japanese Business Studies Conference*, St Louis, Missouri, February 2002.
Dosi, G. (1988) Sources, procedures and microeconomic effects of innovation. *Journal of Economic Literature*, 26(3): 1120–71.
Dussauge, P., Garrette, B. and Mitchell, W. (2000) Learning from competing partners: outcomes and durations of scale and link alliances in Europe, North America and Asia. *Strategic Management Journal*, 21(2): 99–126.
Garvin, D. (1993) Building a learning organization. *Harvard Business Review*, July–August, 78–91.
Grant, R. M. (1996) Toward a knowledge-based theory of the firm. *Strategic Management Journal*, 17(Winter special issue): 109–22.
Gulati, R. (1995) Structure and alliance formation patterns: a longitudinal analysis. *Administrative Science Quarterly*, 40(4): 619–52.
Hamel, G. (1991) Competition for competence and inter-partner learning within international strategic alliances, *Strategic Management Journal*, 12(special issue): 83–103.
Hedberg, B. (1981) How organizations learn and unlearn. In P. C. Nystrom and W. H. Starbuck (eds) *Handbook of Organizational Design*, Vol. 1. New York: Oxford University Press, pp. 5–23.
Hedlund, G. (1994) A model of knowledge management and the N-form corporation. *Strategic Management Journal*, 15(Summer special issue): 73–90.
Hippel, E. von (1994) Sticky information and the locus of problem solving: implications for innovation. *Management Science*, 40(4): 429–39.
Howell, J. and Higgins, C. (1990) Champions of technological innovation. *Administrative Science Quarterly*, 35(2): 317–41.

Huber, G. (1991) Organizational learning: the contributing processes and the literatures. *Organization Science*, 2(1): 88–115.

Inkpen, A. (1999) *Wil-Mor Technologies*. Teaching Case and Teaching Note. London, Canada: Ivey Management Services, University of Western Ontario.

Inkpen, A. C. and Beamish, P. W. (1997) Knowledge, bargaining power, and the instability of international joint ventures. *Academy of Management Review*, 22: 177–202.

Inkpen, A. and Dinur, C. A. (1998) Knowledge management processes and international joint ventures. *Organization Science*, 7(3): 211–20.

Kale, P., Singh, H. and Perlmutter, H. (2000) Learning and protection of proprietary assets in strategic alliances: building relational capital. *Strategic Management Journal*, 21(3): 217–37.

Khanna, T., Gulati, R. and Nohria, N. (1998) The dynamics of learning alliances: competition, cooperation and relative scope. *Strategic Management Journal*, 19(3): 193–210.

Kogut, B. (1988) Joint ventures: theoretical and empirical perspectives. *Strategic Management Journal*, 9: 319–32.

Kogut, B. and Zander, U. (1992) Knowledge of the firm, integration capabilities, and the replication of technology. *Organization Science*, 3: 383–97.

Kogut, B. and Zander, U. (1996) What firms do? Coordination, identity, and learning. *Organization Science*, 7(5): 502–18.

Lane, P. J. and Lubatkin, M. (1998) Relative absorptive capacity and interorganizational learning. *Strategic Management Journal*, 19: 461–77.

Lengnick-Hall, C. A. and Wolff, J. A. (1999) Similarities and contradictions in the core logic of three strategy research streams. *Strategic Management Journal*, 20(12): 1109–32.

Levitt, B. and March, J. G. (1988) Organizational learning. *Annual Review of Sociology*, 14: 319–40.

Lyles, M. A. (1988) Learning among joint venture sophisticated firms. *Management International Review*, 28(special issue): 85–98.

Lyles, M. A. (1994) An analysis of discrimination skills as a process of organizational learning. *The Learning Organization*, 1(1): 23–32.

Lyles, M. A. and Salk, J. E. (1996) Knowledge acquisition from foreign parents in international joint ventures: an empirical examination in the Hungarian context. *Journal of International Business Studies*, 29(2): 154–74.

Maidique, M. A. and Zirger, B. J. (1985) The new product learning cycle. *Research Policy*, 14(6): 299–309.

Makhija, M. V. and Ganesh, U. (1997) The relationship between control and partner learning in learning-related joint ventures. *Organization Science*, 8: 508–27.

Mezirow, J. (1991) *Transformative Dimensions of Adult Learning*. San Francisco, CA: Jossey-Bass.

Mowery, D. C., Oxley, J. E. and Silverman, B. S. (1996) Strategic alliances and interfirm knowledge transfer. *Strategic Management Journal*, 17(special issue): 35–9.

Nelson, R. R. and Winter, S. G. (1982) *An Evolutionary Theory of Economic Change*. Cambridge, MA: Belknap Press.

Nonaka, I. (1994) A dynamic theory of organizational knowledge creation. *Organization Science*, 5(1): 14–37.

Nonaka, I. and Takeuchi, H. (1995) *The Knowledge Creating Company*. New York: Oxford University Press.

Nystrom, P. C. and Starbuck, W. H. (1984) To avoid organizational crises, unlearn. *Organizational Dynamics*, 13: 53–65.
Parkhe, A. (1991) Interfirm diversity, organizational learning, and longevity in global strategic alliances. *Journal of International Business Studies*, 22(4): 579–601.
Parkhe, A. (1998) Building trust in international alliances. *Journal of World Business*, 33(4): 417–37.
Polanyi, M. (1966) *The Tacit Dimension*. Garden City, NY: Doubleday.
Postman, L. and Stark, K. (1965) The role of response set in tests of unlearning. *Journal of Verbal Learning and Verbal Behavior*, 4: 315–22.
Powell, W. W. (1998) Learning from collaboration: knowledge and networks in the biotechnology and pharmaceutical industries. *California Management Review*, 40(3): 228–40.
Powell, W. W., Koput, K. W. and Doerr, L. S. (1996) Interorganizational collaboration and the locus of innovation: networks of learning in biotechnology. *Administrative Science Quarterly*, 41: 116–45.
Preskill, H. and Torres, T. T. (1999) The role of evaluative enquiry in creating learning organizations. In M. Easterby-Smith, J. Burgoyne and L. Araujo (eds) *Organizational Learning and the Learning Organization*. London: Sage.
Ring, P. S. and Van de Ven, A. (1992) Structuring cooperative relationships between organizations. *Strategic Management Journal*, 13: 483–98.
Ring, P. S. and Van de Ven, A. (1994) Developmental processes of cooperative interorganizational relationships. *Academy of Management Review*, 19(1): 90–118.
Schon, D. A. (1983) *The Reflective Practitioner: How Professionals Think in Action*. New York: Basic Books.
Senge, P. M. (1990) *The Fifth Discipline*. New York: Doubleday.
Shaw, R. B. and Perkins, D. N. T. (1991) Teaching organizations to learn. *Organization Development Journal*, 9(4): 1–12.
Simonin, B. L. (1997) The importance of developing collaborative know-how: an empirical test of the learning organization. *Academy of Management Journal*, 40(5): 1150–74.
Simonin, B. L. (1999) Ambiguity and the process of knowledge transfer in strategic alliances. *Strategic Management Journal*, 20(7): 595–623.
Szulanski, G. (1996) Exploring internal stickiness: impediments to the transfer of best practice within the firm. *Strategic Management Journal*, 17(special issue): 27–43.
Teece, D. (2000) Strategies for managing knowledge assets: the role of firm structure and industrial context. *Long Range Planning*, 33(1): 35–54.
Urban, G. L. and Hippel, E. von (1988) Lead user analyses for the development of new industrial products. *Management Science*, 34(5): 569–83.
Winter, S. G. (1987) Knowledge and competence as strategic assets. In D. J. Teece (ed.) *The Competitive Challenge: Strategies for Industrial Innovation and Renewal*. Cambridge, MA: Ballinger.
Zaheer, A., McEvily, B. and Perrone, V. (1998) Does trust matter? Exploring the effects of interorganizational and interpersonal trust on performance. *Organization Science*, 9(2): 141–59.

6

Learning versus Protection in Inter-Firm Alliances: A False Dichotomy

Joanne E. Oxley[1]

Introduction

> ... the pursuit of greater joint value [in inter-firm arrangements] requires the use of governance structures that are *less* efficient from a transaction cost perspective ... strategic and learning gains often increase transaction value while simultaneously increasing transaction costs. (Zajac and Olsen, 1993: 132, 143)
>
> Theorists must adopt long-term efficiency as the criterion, and they must address such variables as innovation, learning and asset redeployability. (Ghoshal and Moran, 1996: 41)
>
> ... a normative use of the TCE [transaction cost economics] framework within organizations only serves to *heighten* the potential conflict, *reduce* the potential for mutual gains, and *limit* the means by which order can be accomplished. (Moran and Ghoshal, 1996: 61)

These quotations reflect ideas frequently voiced by researchers studying strategic alliances and other inter-firm arrangements from a knowledge-based or resource-based perspective: i.e. that alliance participants face a fundamental conflict between the desire to learn from partners (or to innovate within the alliance) and the need to protect themselves from potential opportunistic behaviour by those same alliance partners. The hazards of partner-firm opportunism include the risk of technology leakage, or unauthorized use of the firm's technology or know-how (Teece, 1986; Hamel, Doz and Prahalad, 1989; Oxley, 1997), and *ex post* extraction of rents generated through irreversible relationship-specific investments (so-called 'hold-up' hazards).

In this chapter, I argue that this dichotomy, between learning and protection in alliances, is in fact a *false* dichotomy, based on a 'straw man' version of transaction cost economics (TCE). Implicit in these characterizations of alliance organization is the assumption that transaction cost economics relies on a narrow, short-term view of efficiency and that 'hierarchy' is synonymous with increased monitoring, which in turn constrains learning. Critics overlook the role of *credible commitments* (e.g. irreversible investments) in enhancing learning opportunities *and* protection, through the lowering of incentives to engage in opportunistic behaviour.

My defence of the transaction cost framework is not meant to imply that TCE alone explains all important facets of inter-firm alliance behaviour. Indeed, I argue that the resource-based view of alliance organization contributes substantially to our understanding of inter-organizational arrangements by addressing a set of questions that has largely been ignored by researchers in TCE. The resource-based view conceives of the firm as a collection of 'sticky' and hard-to-imitate assets (Penrose, 1959; Wernerfelt, 1984; Conner, 1991). Research in this tradition has analysed the processes by which rents can be captured through protection and deployment of idiosyncratic resources (Barney, 1986; Dierickx and Cool, 1989). More recently the dynamic process of change in capabilities underpinning firm-specific resources has been explored (Teece and Pisano, 1994; Teece, Pisano and Shuen, 1997), as well as the role of inter-firm alliances in the acquisition of new capabilities through organizational learning (e.g. Teece and Pisano, 1994: 545). The RBV therefore focuses attention on questions relating to the nature of capabilities developed and acquired within alliances, to the impact of external environmental shocks on alliance activities, and to the design of organizational processes necessary to support inter-firm learning. The comparative lens of TCE nonetheless remains an indispensable tool in determining the relative efficacy of different forms of organization for learning or capabilities development.

I argue that it is vital for research in alliances that we pursue a *combined* approach, melding the insights of transaction cost economics with those of the resource-based view. Only in this way can we build and test models that explicitly acknowledge the presence of significant heterogeneity in the capabilities brought together within alliances. This is important because such heterogeneity has implications for the type of interaction necessary to support learning and other performance-enhancing activities, which in turn has implications for the hazards facing partner firms, and the types of governance structures necessary to support cooperation within an alliance. Following discussion of the complementarities between the resource-based view and TCE, I make suggestions for future research to further our understanding of learning and protection within inter-firm alliances, based on this premise.

Transaction cost economics as 'straw man'

In his introduction to the *Organization Science* special issue devoted to the resource-based theory of the firm, Barney (1996) notes that the controversy raised by this emerging theory is unsurprising, given the existence of 'a widely accepted theory of the firm' within the organizational economics literature, dating back to Coase (1937) and Williamson (1975). Barney suggests that, in these circumstances, any newly proposed theory must explain not only the phenomenon of interest, but also why a new theory needs to be developed at all. Such an explanation 'necessarily involves discussing the limitations and weaknesses of traditional transaction cost theories of the firm' (Barney, 1996: 469). This is undoubtedly true, and the critiques and counter-arguments of such a debate play a valuable role in the development of organization theory (and of social science more generally). However, there is an attendant danger that, in trying to carve out a space for new theory, one ends up attacking a straw man version of the existing framework and raising false areas of conflict, where in fact a combined approach may be more fruitful.

The following discussion highlights aspects of TCE that have been reduced to such straw man proportions in some research on inter-firm arrangements by scholars adopting a knowledge-based or resource-based view of the firm. The focus on research into inter-firm arrangements is motivated by several observations. First, some of the more distorted characterizations of TCE can be found in this literature, and the research agenda has at times been driven by the resulting false dichotomy drawn between learning and protection in alliances. Second, alliances by definition involve a relationship between entities with separate identities, communities and cultures. We can therefore sidestep some of the debate regarding social identity as a determinant of firm boundaries (Kogut and Zander, 1992, 1996). Furthermore, the broader debate on the theory of the firm is quite complex and wide-ranging, and goes well beyond the scope of a single paper.[2] Observations on this debate are therefore limited to instances where direct connections to the alliance literature are apparent.

Before turning to the critiques of TCE as a lens for analysing learning in alliances, a word should be said on what we mean by learning here, as there is much variety in the theoretical interpretation and empirical operationalization of learning in the alliance literature. One common interpretation of alliance learning is the transfer of skills and capabilities from one partner firm to another (Hamel, 1991; Kogut, 1988; Sobrero and Roberts, 2001). In this case, alliances serve as vehicles for partner firms to internalize capabilities that are difficult to acquire through alternative channels. Another interpretation relates to the more general lessons a firm may learn about how to design and manage alliances, so increasing the effectiveness of subsequent alliances (Lyles, 1988; Simonin, 1997) and perhaps creating a unique 'relational advantage' (Dyer and Singh, 1998). Alternatively, a firm may learn about the prospects for success of a particular project, or about the behavioural tendencies of a particular alliance partner (Kogut, 1991; Mody, 1993). Finally, learning may refer to the pooling of knowledge-based resources by alliance partners in order to create entirely new knowledge, for example within an R&D alliance or consortium (Inkpen and Dinur, 1998; Olk and Young, 1997; Sampson, 2003).

Despite this diversity in interpretations, one common requirement for learning in alliances is the need for both partners to commit significant resources (including open access to information) to the venture, and to be responsive to changing resource needs as circumstances develop. The question then becomes; does moving towards hierarchy in an alliance impede or facilitate such processes? This is the crux of the issue contested within the alliance literature.

Efficiency versus learning

Transaction cost economics has traditionally been concerned with the make-or-buy decision in intermediate goods markets. Firms are expected to structure transactions in such a way as to minimize the sum of production and transaction costs, by designing governance structures that support any necessary investments in relationship-specific assets, while guarding against the hazards of opportunism by transaction partners. This focus on governance structure design has sometimes been characterized as an exercise in the pursuit of short-term efficiency, leaving little room for notions such as 'strategic and learning gains' (Zajac and Olsen, 1993) or 'long-term learning enhancing outcomes' (Sobrero and Roberts, 2001). Does this mean that transaction cost economics is an inappropriate lens for addressing issues related to learning? I suggest not.

Consider the following example. Two firms are faced with outsourcing decisions for the manufacture of a new component. Firm A views the component as a modular innovation that can most economically be produced in a large-scale plant equipped with relatively general-purpose equipment. There are existing producers available, operating at an efficient scale, and since little relationship-specific investment is foreseen on either side of the transaction, a simple supply contract is adopted.

Firm B is considering the outsourcing of a similar component, but foresees a learning process occurring, with fine-tuning in the way that the component interfaces with other parts of the product. This learning process is expected to require frequent interactions among the various functional areas involved in the manufacture of the component and other parts of the final product, leading to development of considerable relationship-specific know-how. Firm B, taking the long-term contracting perspective of transaction cost economics, predicts that a standalone manufacturer with only a short-term contract may be unwilling to invest in the development of relationship-specific know-how that could become subject to hold-up in the future (since the value of the investments would be lost in the event the agreement is terminated). In anticipation of these problems, Firm B proposes a long-term contract or some other type of strategic alliance, or perhaps even an equity-sharing arrangement such as a joint venture. These latter arrangements represent 'hybrid' governance structures that (as argued in more detail below) offer greater assurance of continued cooperation and thus support greater investment in relationship-specific assets (Oxley, 1997).

Among these two scenarios, which is the more 'efficient' from a transaction cost perspective? Which results in the best 'learning-enhancing outcomes?' Which maximizes 'transaction value'? Zajac and Olsen (1993) clearly associate transaction cost economizing only with Firm A's decision to pursue a narrowly defined transaction and a short-term contract, since they argue that 'interorganizational strategies having greater joint value will typically require the use of *less* efficient (from a transaction cost perspective) governance structures' (Zajac and Olsen, 1993: 138). This in turn implies that only transactions with very low levels of asset specificity are correctly viewed as efficient from a transaction cost perspective. This need not, of course be the case, since transaction cost economizing implies minimizing the sum of transaction *plus* production costs, for a given value-creating transaction (Riordan and Williamson, 1985; Williamson, 1985: 92–3).

The normative implication of transaction cost logic is thus that firms should look forward and identify those situations where there is the likely potential for valuable learning (or other requirements for bilateral adaptation), and craft governance structures that facilitate the necessary development of relationship-specific know-how and other assets in an efficient (i.e. transaction cost economizing) way.[3] Of course, foresight in these matters is imperfect at best, and learning is often an 'emergent' process, as emphasized in the literature on organizational learning (e.g. Huber, 1991). However, I would submit that lack of precision in forecasting does not warrant an abandonment of transaction cost economizing in inter-firm arrangements.

Building on ideas regarding the *dynamic* nature of learning, Ghoshal and Moran (1996) suggest that managers following the logic of TCE are inevitably concerned with only static cost efficiencies and will therefore forego many opportunities for innovation and learning. This is so, according to Ghoshal and Moran (1996: 34), because many such learning opportunities are efficient only in a dynamic sense and therefore require firms to 'hold off' market forces, at least temporarily. Certainly it is true that, absent compelling reasons to do otherwise, the transaction cost logic implies that the scope of transactions should be restricted so that simple (contractual) governance structures will suffice in inter-firm relationships (Oxley, 1997). However, where investments in learning and other relationship specific assets are expected to yield high future returns, this will be factored into the (efficient) governance decision (Williamson, 1996b: 52). Thus, to argue that firms adopting TCE will emphasize efficiency-seeking at the expense of learning or innovation opportunities requires a very myopic view of firm behaviour, such that managers are unable to predict with any confidence which relationships are likely to provide learning opportunities. And in the face of such myopia, Ghoshal and Moran's (1996) argument implies that the only strategy to enhance learning and dynamic efficiency is continually to expand the scope and flexibility of transactions with *all* transaction partners, in the hope that potential learning benefits will outweigh ongoing production and governance cost penalties. But this begs a question: how should relationships be bounded or the scope of the firm determined? Ghoshal and Moran (1996) are silent on this issue.

Monitoring, safeguards and credible commitments

The second aspect of the straw man version of TCE prevalent in writing on inter-organizational relationships is the identification of 'hierarchy' with bureaucratic apparatus and 'big brother' monitoring mechanisms that impede innovation and learning. Moran and Ghoshal (1996: 61), for example, argue that adoption of a TCE perspective serves to 'heighten conflict, reduce potential for mutual gains, and limit means by which order can be accomplished'. Drawing on arguments from psychology, and citing research on the impact of surveillance on motivation at work and at play (Enzle and Anderson, 1993; Lepper and Greene, 1975; Strickland, 1958), they suggest that hierarchical controls create a 'negative feeling for the entity' (Ghoshal and Moran, 1996: 23). Their focus on surveillance is consistent with Ghoshal and Moran's (1996) view of hierarchy, which explicitly identifies hierarchical controls with surveillance and fiat. However, as argued below, this is a somewhat misleading characterization of the nature of hierarchical governance as conceived by transaction cost economists. By elaborating on the governance mechanisms found in inter-organizational alliances I suggest that moving to more hierarchical control generally has the effect of *increasing* rather than decreasing cooperation, and enhancing the ability to learn from alliance partners.

The transaction cost view of an inter-organizational alliance is that of a hybrid governance form, lying between the polar forms of market and hierarchy (Williamson, 1991). While markets are argued to be superior in terms of high-powered incentives and autonomous or 'Hayekian' adaptability (e.g. unilaterally responding to a change in price or other demand signal), internal organization promotes the ability to effect *bilateral* adaptation, or a coordinated response to disturbances that result in a potential conflict of interests between the transacting parties. Where significant uncertainty is present (in conjunction with relationship-specific investments), autonomous adaptation can have perverse effects and so internal organization is favoured. Alliances meld some of the governance attributes of these two polar governance forms, and thus lie between market and hierarchy in terms of their ability to support the two different types of adaptation.

Given the myriad interorganizational arrangements employed by firms today, it is of course apparent that all alliances are not created equal in terms of their governance attributes. Interorganizational arrangements that are gathered under the rubric of alliances include organizational forms as diverse as technical training and start-up assistance arrangements, production, assembly and buy-back agreements, patent or know-how licensing, franchising, management or marketing service agreements, non-equity cooperative agreements in R&D, development or co-production, and equity joint ventures (Contractor and Lorange, 1988). Conceptually, these alliances may nonetheless be viewed as lying along a 'market-hierarchy continuum' of governance forms, with some forms exhibiting governance features closer to those of the market and some approaching the governance properties of firms. The key dimensions that distinguish such a continuum are the degree to which the partners' fortunes are tied together (i.e. incentive alignment), the administrative controls available, and the nature of the legal supports (Oxley, 1997).[4]

One of the key features of TCE is its focus on 'discrete structural alternatives' (Williamson, 1991). The idea is that, although there is significant variety in the governance arrangements that firms adopt, it is useful to identify discrete categories of governance, within which there is admittedly variation, but which nonetheless represent tight clusters, in governance terms. This allows careful comparative analysis of the incentive and adaptive features of different governance modes. At the simplest level, for inter-firm alliances, the market-hierarchy continuum can be reduced to a choice between contracts and equity-based alliances (joint ventures).

In 'learning alliances' (e.g., technology sharing or R&D alliances) where success demands that both firms are willing to commit substantial firm-specific resources and to jointly adapt resource commitments to changing needs, contractual governance is problematic, for the reasons discussed earlier – potential leakage of proprietary information and/or hold-up problems, and difficulties in settling complex disputes in court. One response to these contracting problems is for firms to limit information sharing, by reducing the transparency of their operations or employing elaborate gate-keeping mechanisms (see, for example, Hamel *et al.*, 1989), or to write ever-more complex contracts defining in ever-increasing detail the rights and responsibilities of the partner firms.

This response, often associated with a transaction cost approach to alliance governance[5] appears to exemplify the very dichotomy – between protection and learning – that I contend is false. And certainly, if it were the case that contracts exhausted the governance choices available to alliance participants, then managers would indeed face a troubling dilemma: incomplete contracts could be remedied only by progressively more elaborated contractual terms, more vigilant monitoring of alliance partner activities, and narrowing of the transfers of technology and know-how among partners. Such a narrowing would inevitably reduce opportunities for learning. However, in reality, transaction cost economics suggests an alternative response, which is to look to alternative methods of governing the relationship or transaction that enhance bilateral adaptation – in this simple illustration, by moving to an equity joint venture.

The equity joint venture is perhaps the archetypal inter-firm organization, and as such has been the focus of much prior research (e.g. Geringer and Hebert, 1989; Gomes-Casseres, 1989; Harrigan, 1986; Hennart, 1991; Killing, 1983; Pisano, Russo and Teece, 1988). In governance terms, the shared equity in the new venture creates a hostage exchange, or exchange of *credible commitments* between the partners to the alliance. Because the value of the joint venture depends critically on continued operation, each firm effectively posts a bond equal to its equity share (the value of which is at best only partially redeemable should the venture terminate prematurely). And since the ongoing returns to each partner are also based on the profits of the venture as a whole, the incentives of the partners to strive for jointly optimal outcomes are enhanced.

In addition to aligning incentives in this way, a joint venture structure improves the *ability* of alliance partners to adapt resource commitments in a coordinated manner as changing circumstances demand. Along with the pooling of financial resources, joint venture partners also pool managerial control by having a board of directors that typically includes members from partner firms in proportion to equity holdings. This provides a direct communication link with senior management of the parent companies, and is a conduit for strategic directives. Note, however, that in contrast to in-house activities (or an arm's-length arrangement), these directives are translated into action via negotiation and compromise between the parent companies. Indeed, the right of veto over strategic decisions is often explicitly incorporated in the formal agreement accompanying the creation of a joint venture (Geringer and Hebert, 1989; Killing, 1983). This ensures that the interests of both partners are recognized in the adaptive moves of the venture.

Together these features of the equity joint venture structure create a strong bilateral dependence, and the incentives to behave opportunistically, i.e., to cheat on the agreement, are thus reduced (Pisano, 1989; Williamson, 1983). By making credible commitments, each firm benefits from the greater confidence engendered in the partner, which promotes a willingness to share information and know-how more freely on both sides. The ability to coordinate adaptations is also facilitated through the administrative structure of the joint venture. It is in this sense that greater hierarchical control in an alliance promotes learning, contrary to Ghoshal and Moran's (1996) characterization.

Many researchers studying alliances reject what they see as TCE's over-emphasis on the role of contractual features of interfirm agreements and argue that learning and protection can only be achieved simultaneously in the context of a relationship that fosters the development of 'trust' and norms of reciprocity (Barney and Hansen, 1995; Casson and Nicholas, 1989; Powell, 1990; Shane, 1994).[6] Thus, in recent theoretical and empirical studies of alliance performance, researchers have attempted to separate the impact of trust from that of contractual governance (e.g. Kale, Singh and Perlmutter, 2000; Poppo and Zenger, 2002). I would argue, however, that while this research has focused attention on informal aspects of governance that have been hitherto under-researched, it is inappropriate to characterize trust as something separate from contractual governance; trust is more appropriately viewed as an *outcome* of appropriate governance choice which includes, but is not limited to, the choice of contractual terms.

This focus on broader features of alliance governance is also consistent with Williamson's (1991) argument, that hybrid governance modes are supported by neoclassical or relational contracting: the 'highly adjustable framework' of neoclassical contracting explicitly supports the development of norms of reciprocity (Llewellyn, 1931, cited in Williamson, 1991: 272; see also Macneil, 1978). The notion is that small disturbances, or movements 'off the contract curve', will not be renegotiated on a case-by-case basis, but rather will be absorbed by alliance partners. This accommodation is supported by the understanding that such disturbances will either balance out between the partners over the medium- to long-term or, if unbalanced differences persist, adjustments will subsequently be made through a previously designated adjustment process, or possibly through arbitration.

In part, these norms of reciprocity (and the implied trust between the partner firms) that emerge in neoclassical are supported by the hostage exchanges implicit in a bilateral arrangement. Such norms will be further enhanced where partners have multiple ongoing cooperative ventures together since, in this case, the pay-off to opportunism within each individual alliance is lower, because of the risk that continued gains from cooperation in all of the alliances will be withdrawn if opportunism is detected (Gulati, 1995a; Kogut, 1989).

In addition to attenuating moral hazard problems, repeat alliances may also act as a screening device, reducing adverse selection problems in partner choice. Improved information is developed over the course of cooperative projects regarding a partner's propensity to engage in opportunistic behaviour (Balakrishnan and Koza, 1993). With greater confidence in the ability to work things out with the partner, without recourse to formal dispute resolution mechanisms, the need for hierarchical controls in subsequent alliances is reduced. Similarly, firms embedded in a dense relational network will be better able to screen potential alliance partners (at least when drawn from the same network) and can themselves effectively commit to desist from opportunistic behaviour – evidence of such behaviour will predictably damage the firm's reputation, and the negative consequences may spill over into its other relationships within the network, so exacting a high cost.

The alliances just described, involving hostage exchange arrangements and relational contracts, are, as suggested by advocates of the relational view of alliances, generally superior to a simple tightly-specified contract for promoting learning, since they support coordinated adaptation by the partner firms (Dyer and Singh, 1998). At the same time, they do not necessitate all of the additional start-up and bureaucratic costs associated with an equity joint venture.[7] However, looking at these alliances through the lens of TCE also suggests important limits inherent to informal alliances: since the partners retain a greater degree of legal and structural autonomy than is the case in a joint venture, the zone of 'self-enforcement' of the contract is lower than is the case for a joint venture (all else equal). Furthermore, for first-time collaborators – who cannot rely on the opportunism-attenuating effects of related investments or reputation – the greater commitment and incentive alignment properties of an equity joint venture will be particularly attractive. This brings us to the final criticism of the transaction cost framework salient in research on inter-firm alliances which must be addressed – its alleged excessive focus on opportunism.

Tacit know-how, learning and opportunism

Many critiques of the TCE paradigm are based on the premise that, while plausible in certain circumstances, TCE relies on an overly legalistic view of organization and reflects an unnecessarily jaundiced view of human nature (e.g. Conner and Prahalad, 1996; Madhok, 1996, 1998). Indeed, the argument has been advanced elsewhere that opportunism is quite unnecessary to explain variety in the way that inter-firm alliances are structured. For the choice between a contract-based alliance and an equity joint venture, for example, Kogut (1988) argues that joint ventures are the preferred vehicle by which tacit knowledge is transferred, and contractual modes are ruled out, 'not because of market failure or high transaction costs as defined by Williamson and others, but rather because the very knowledge being transferred is organizationally embedded' (Kogut, 1988).

This argument, which does not rely on opportunism of the partners in an alliance, is apparently at odds with the transaction cost framework described so far. However, I would suggest that the two approaches are in fact complementary. Kogut and Zander's (1996) knowledge-based theory offers many important insights into the process of (and obstacles to) know-how transfer within and between firms that have been overlooked within TCE, but it falls short of explaining the choice among different governance alternatives for inter-firm alliances.

According to Kogut and Zander (1996: 503), the way that a firm should be understood is as a 'social community specializing in the speed and efficiency of creation and transfer of knowledge'. Knowledge is more effectively created and transferred within a firm than through the market because firms provide 'a sense of community by which discourse, coordination and learning are structured by identity'. Kogut and Zander (1992, 1996) draw several inferences from this insight that have relevance to the choice of organizational form for know-how transfer. First, the most important factor in a make-or-buy decision is the differential capabilities of the firm and its suppliers (1992: 394). Second, know-how transfer requires 'frequently interaction within small groups, often through the development of a unique language or code[8]... It is the sharing of a common stock of knowledge, both technical and organizational, that facilitates the transfer of knowledge within groups' (1992: 389). Furthermore, communication *across* groups (for instance across functional areas within the firm, or across firm boundaries) is facilitated by certain individuals occupying pivotal roles as 'boundary spanners' (1992: 389). Finally, identity facilitates the development of 'convergent expectations' (1996: 511), through which coordination is achieved.

When we look at the choice between an equity joint venture and a contract-based or informal alliance for transferring know-how, the limits of this argument come into sharp focus. Here, the partners cannot be assumed to share a 'unique language or code' in either case since they are, by definition, autonomous firms. Similarly, there is no a priori reason to expect that individuals in each organization will identify with the other partner in a joint venture any more than is the case in some other type of alliance. Still, one appealing implication of Kogut and Zander's (1992, 1996) argument is that difficulty in transferring tacit know-how requires that personnel are co-located for a nontrivial time period to facilitate learning by doing, experimentation, demonstration, feedback – processes that are necessary for effective organizational learning (Garvin, 1993). Thus, 'to the extent that close integration within a supplier or buyer network is required, long-term relationships embed future transactions within a learned and shared code' (Kogut and Zander, 1992: 390). Support for this premise can be found in previous studies of the direct costs of technology transfer within and between multinational firms (Teece, 1977, 1981), as well as in studies indicating that inter-partner knowledge transfer is greater in equity joint ventures than in contract-based alliances (Mowery, Oxley and Silverman, 1996).

The question nonetheless remains; why is the necessary co-location and/or long-term interaction of personnel most effectively achieved within an *equity joint venture* (as argued by Kogut, 1988)? After all, absent opportunism, the two firms could simply write a simple general clause contract agreeing to pool personnel, perhaps in a separate facility (as is the case in many joint ventures), share know-how, and distribute benefits based on some pre-agreed sharing rule (Foss, 1996). In reality, it is apparent that such an arrangement is fraught with hazards related to the misappropriation of know-how, or hold-up in the face of relationship-specific investments. And it is precisely the governance features of the equity sharing and joint managerial control in a joint venture that mitigate these hazards and imbue the relationship with the confidence necessary for the commitment of resources and know-how sharing (i.e. learning).[9]

Capabilities, competition and alliance organization

The above argument implies that Kogut and Zander's (1992, 1996) knowledge-based theory does not contain a full set of *sufficient* conditions for an explanation of organizational (and governance) structures in inter-firm alliances.[10] The theory nonetheless contributes greatly to our understanding of one of the central *necessary* conditions for alliance activity in many contexts, i.e., the organizational challenges associated with learning (with their roots in the nature of know-how) and the process of development of the stock of knowledge which, in large part, defines 'what firms do' (Kogut and Zander, 1996). Inkpen (1996) and Makhija and Ganesh (1997) have further elaborated the challenges encountered in learning from alliance partners, and the implications that this has for organizational interfaces to facilitate knowledge transfer. The organizational requirements of innovation and learning remains an underdeveloped area within TCE, reflecting the early focus on make-or-buy decisions in technologically mature industries. While the oversight has been partially corrected in recent years (e.g. Sampson, 2003), Kogut and Zander (1996) and other researchers adopting a knowledge- or resource-based view of the firm probe more deeply into how a firms knowledge base develops over time, and how this translates into sustained competitive advantage.

This is but one example of the areas where the resource-based view of the firm provides important insights into the organization of inter-firm alliances. The resource-based view brings three fundamental insights to the analysis of inter-firm alliances, prompting researchers to ask questions that hitherto have been neglected by transaction cost economists. These insights are (1) that there is persistent and economically significant heterogeneity in firm capabilities, (2) that this heterogeneity – particularly as it relates to 'core competence' (Prahalad and Hamel, 1990) – drives the differential ability of firms to generate rents in end-product markets; and (3) that the pay-offs from cooperative and competitive behaviour within an alliance will depend significantly on the relative capabilities and market positions of partner firms (Khanna, Gulati and Nohria, 1998). Below, I propose a direction for future research on inter-firm alliances, motivated by these key insights, capitalizing on the complementarities between TCE and the resource-based view of the firm.

Transaction cost economists have paid little attention to the heterogeneity of potential partners' capabilities, *ex ante,* except to the extent that the lack of a 'thick' market in the needed assets leads to small-numbers problems when structuring the transaction. The decision to collaborate in the TCE framework turns primarily on the nature of the assets to be combined and the resulting contractual hazards – as discussed above, where assets are particularly idiosyncratic and hazards are severe, the transaction cost logic suggests that organizing the activity within a single firm is preferred (either by acquiring the assets in question, or by developing them in-house). However, as research in the resource-based view stresses, it is precisely these idiosyncratic firm-specific assets that may be difficult to imitate (Barney, 1991; Teece, 1986) and that also may not be alienable from their organizational context (Kogut and Zander, 1992), at least in the short to medium term. This means that there may be circumstances where an alliance is the *only* way to bring together a particular combination of idiosyncratic resources that is expected to generate significant rents. This also implies that the choice of *who* to partner with becomes a nontrivial issue in the presence of heterogeneous capabilities.[11]

Bringing together the insights from the resource-based view and TCE holds particular promise for illuminating the implications for alliance organization of competition in technology and product markets. Consider the notion of 'core competence' (Prahalad and Hamel, 1990). By definition, core competences represent investments in highly idiosyncratic assets (generating correspondingly high rents), and their continued value to the firm requires that competitors (or potential competitors) cannot easily imitate these resources. Efforts to leverage such assets through alliances are arguably fraught with hazards. First, the hold-up problem is particularly severe because of the large rents at stake. Second, if achievement of the alliance objectives necessitates pooling of firm-specific capabilities, or knowledge transfer among partner firms, then the inimitability of the core competence may be compromised.

Interestingly, despite the recognition that competition for technological leadership may lead to the phenomenon of a 'learning race' between alliance partners (e.g. Hamel, 1991; Khanna, Gulati and Nohria, 1998), little is yet known about the implications of this phenomenon for the organization or performance of alliances. One recent study (Mowery, Oxley and Silverman, 2001) suggests that when alliance partners have overlapping core capabilities they may be less willing to reduce activity in (or otherwise cede control over) these core technological domains and that this reduces the extent of complementary specialization achieved within the alliance. This preliminary empirical result suggests an interesting opportunity for future research: i.e. to explore more thoroughly the interaction between (1) characteristics of alliance partner firms (in terms of capabilities and product market competition); (2) the nature and scope of activities undertaken within the alliance; and (3) attributes of the governance structure needed to achieve the alliance goals.

Some preliminary thoughts on the issue of alliance scope and governance may illuminate the power of the combined RBV/TCE approach for issues involving the interaction of firm capabilities, strategic position, and organization.[12] In their commentary on the rise of international alliances in the 1980s, Hamel, Doz and Prahalad (1989: 135) suggested that ideal alliance partners were firms whose 'strategic goals converge while their competitive goals diverge'. The rationale behind this prescription is that if alliance partners are competitors in end product markets (i.e. if their competitive goals 'converge') then each will be so intent on internalizing the other's knowledge, at the same time as limiting access to their own proprietary skills (i.e. acting opportunistically), that the goal of the alliance will be thwarted. However, casual observation reveals that collaboration among direct product-market competitors is commonplace, particularly in alliances for development of new technology (e.g. R&D alliances). The resource-based view provides a rationale for why such alliances may be beneficial, since it is likely to be that firms active in the same end-product markets have the most to learn from each other, and that, absent competitive considerations, such firms could most profitably combine their capabilities in developing new technologies that will be future sources of sustainable, rent-generating, competitive advantage. However, as the prescriptions of Hamel *et al.* (1989) suggest, mitigating the hazards of opportunistic behaviour in alliances involving competitors is particularly challenging. The 'traditional' response to such hazards from a TCE perspective is of course the adoption of a more hierarchical governance structure, and indeed one conjecture that derives straightforwardly from a combined RBV/TCE perspective on this question is that when direct competitors form an alliance they are likely to adopt a more hierarchical governance structure, all else being equal.[13]

It may be, however, that where end-product market competition is particularly intense, it is not possible to sufficiently attenuate incentives for opportunistic behaviour even within an equity joint venture, to support large-scale knowledge sharing. One alternative is, of course, to forego cooperation in such instances. However, a further alternative, which has been suggested in prior TCE research (e.g. Oxley, 1999), but which has yet to be fully explicated or empirically tested, is to modify the type of activities undertaken within the alliance, i.e. to reduce alliance 'scope'. In particular, if the activities performed within the alliance are designed in such a way as to allow the technology or product development project to the effectively 'modularized' to reduce the need for extensive knowledge sharing, then firms may be able to reap at least some of the benefits of cooperation with competitors while reducing transaction costs associated with the elevated hazards of opportunism. Thus a complete analysis of the decisions underlying alliance governance in the presence of heterogeneous capabilities and product-market competition requires simultaneous analysis of the antecedents of alliance scope and governance. This represents just one example of the many interesting and challenging questions that can be effectively tackled with a combined RBV/TCE approach.

Conclusion

In arguing for continued and enhanced conversation between scholars adopting the resource-based view and those operating in related research areas such as strategic group theory, organizational economics and industrial organization, Mahoney and Pandian (1992) suggest that:

> [while] a morality play of the virtuous resource-based theorists doing battle against the misguided strategic group theorists and industrial organization analysts may provide a crusading faith for the young and naive, a more balanced view... is needed. Intellectual isolating mechanisms which artificially reduce the trading of ideas are not best for the strategy field as a whole. (p. 374)

The arguments in this paper echo this sentiment with respect to transaction cost economics, highlighting areas where efforts to distinguish resource-based analysis of inter-firm alliances have on occasion led researchers to raise a straw man version of transaction cost theory. This is unfortunate for at least two reasons. First, a distorted view of TCE is disseminated in the strategy literature, potentially dissuading scholars from adopting what continues to be a useful lens for analysing a broad range of issues relevant to firm strategy (Williamson, 1991, 1996a). Second, opportunities for research working out of a combined resource-based/transaction cost perspective are overlooked; opportunities which may prove to be more fruitful in increasing our understanding of the role of inter-firm alliances in firm strategy.[14]

Despite some important progress in research into inter-firm alliances from the perspectives of TCE *and* the resource-based view of the firm (as well as other disciplines) there remain significant areas of incomplete understanding. This represents a challenge and an opportunity to conduct research on a phenomenon that has both great importance in itself as well as having implications for broader theoretical debates on the theory of the firm. The overall goal of such a research agenda, 'writ large', is systematically to analyse how firms choose to enter alliances, with whom and for what, and how they organize to maximize benefits and contribute in the 'continuing search for rent' that is the essence of strategy (Bowman, 1974: 47, cited in Mahoney and Pandian, 1992). As suggested in the discussion of alliance scope and governance above, complete answers to these questions require a more fundamental understanding of the nature of firm capabilities and learning, as well as the basic conflicts faced by alliance partners and the role of governance in managing these conflicts. We should look forward with anticipation to the results of increased collaboration between researchers from the resource-based view and from TCE as they tackle these and related issues.

Notes

1. I would like to thank Kathleen Conner, Michael Lawless, Julia Liebeskind, Will Mitchell, Joe Mahoney, Oliver Williamson and Ed Zajac for comments on earlier drafts of this chapter. The usual disclaimers apply.
2. The ideas proposed by Conner and Prahalad (1996), for example, are not discussed here (see Foss, 1996, for a useful critique). While of considerable interest and importance, the emphasis in Conner and Prahalad's work on the role of the employment relationship in shaping the way that knowledge is used within firms precludes straightforward application to inter-firm arrangements (which often do not involve changes in employment relations).

3. Where such opportunities are not foreseen, and general-purpose assets are predicted to be sufficient to support the transaction then, indeed, a narrowly defined transaction and simple contractual governance are advocated.
4. This conceptualization of alliances as lying on a market-hierarchy continuum is not uncontroversial. Critics include sociologists emphasizing the embeddedness of all economic exchange in social and cultural forces (e.g. Granovetter, 1985), those who argue that all forms of exchange involve a complex intermingling of hierarchy and markets (e.g. Eccles, 1985), and those identifying the 'network form' as a distinct organizational phenomenon unconnected to pure market or hierarchical forms (Powell, 1990).
5. Madhok (1998), for example, distinguishes between 'Type I TC' (transaction costs) associated with transaction cost economics and 'Type II TC' associated with knowledge-based constraints on organization. He then suggests that, 'the resolution of differences between firms can be attempted through further strictures on firms' behavior and actions through contractual safeguards (Type I TC). On the other hand, another avenue for resolving differences is through Type II TC ... oriented towards 'education' in the form of cognitive convergence in the pursuit of value' (1998, pp. 18–19).
6. The performance-enhancing benefits of such alliances are thought to be particularly effective when the partner firms are embedded in a network of relationships (Granovetter, 1985; Gulati, 1995b).
7. The bureaucratic costs referred to here include the direct and indirect costs of operating formal dispute resolution procedures. In addition there are costs associated with long communication paths through the joint venture's management structure and board of directors, which are necessary for agreement on major adaptive changes in the venture's operations.
8. A similar idea is advanced in Kogut and Zander, 1996: 'communication is characterized by discourse based on rich codes and classifications, and learning is situated ... ' (1996: 511).
9. Poppo and Zenger (1998) find evidence consistent with this argument in their study of governance performance in make-or-buy decisions in information services. They nonetheless note that empirically differentiating transaction cost and knowledge-based explanations of firm boundaries is difficult.
10. Foss (1996) draws a similar conclusion from his more general analysis of knowledge-based approaches to the theory of the firm.
11. Work has begun here, combining the RBV's focus on capabilities and resources with issues of reputation, hostage exchange and behavioural screening from transaction cost economics, along with sociology's emphasis on the 'embeddedness' aspects of networks that impact alliance partner choice (e.g. Gulati, 1995b; Mowery, Oxley and Silverman, 1998).
12. Other recent examples of the joint consideration of firms' strategic positions, capabilities and governance choices include Ghosh and John (1999), Nickerson, Hamilton and Wada (2001).
13. There is already some empirical evidence that the extent of overlap among allying firms' capabilities has implications for the appropriate choice of governance structure (Sampson, 2003).
14. Others who have called for a synthesis of the resource-based view and transaction cost economics in strategy research, and who have made strides in that direction, include Poppo and Zenger (1998) and Liebeskind (1996).

References

Balakrishnan, S. and Koza, M. P. (1993) Information asymmetry, adverse selection and joint ventures. *Journal of Economic Behavior and Organization*, 20: 99–117.

Barney, J. B. (1986). Strategic factor markets: expectations, luck and business strategy. *Management Science*, 32(10): 1231–41.

Barney, J. B. (1991) Firm resources and sustained competitive advantage. *Journal of Management*, 17(1): 99–120.

Barney, J. B. (1996) The resource-based theory of the firm. *Organization Science*, 7(5): 469.

Barney, J. B. and Hansen, M. H. (1995) Trustworthiness as a source of competitive advantage. *Journal of Management*, 17: 99–120.

Bowman, E. H. (1974) Epistemology, corporate strategy and academe. *Sloan Management Review*, 15(2): 35–50.

Casson, M. and Nicholas, S. (1989) Economics of trust: explaining differences in organizational structure between the United States and Japan. *Discussion Papers in Economics*, 217, University of Reading.

Coase, R. H. (1937) The nature of the firm. *Economica*, 4: 386–405.

Conner, K. R. (1991) A historical comparison of resource-based theory and five schools of thought within industrial organization economics: Do we have a new theory of the firm? *Journal of Management*, 17: 121–54.

Conner, K. R. and Prahalad, C. K. (1996) A resource-based theory of the firm: Knowledge versus opportunism. *Organization Science*, 7(5): 477–501.

Contractor, F. J. and Lorange, P (eds) (1988) *Cooperative Strategies in International Business*, Lexington, MA: Lexington Books.

Dierickx, I. and Cool, K. (1989) Asset stock accumulation and sustainability of competitive advantage. *Management Science*, 35(12): 1504–14.

Dyer, J. H. and Singh, H. (1998) The relational view: cooperative strategy and sources of interorganizational competitive advantage. *Academy of Management Review*, 23(4): 660–79.

Eccles, R. (1985) *The Transfer Pricing Problem: A Theory for Practice*. Lexington, MA: Lexington Books.

Enzle, M. E. and Anderson S. C. (1993) Surveillant intentions and intrinsic motivation. *Journal of Personality and Social Psychology*, 64: 257–66.

Foss, N. J. (1996) Knowledge-based approaches to the theory of the firm: some critical comments. *Organization Science*, 7(5): 470–6.

Garvin, D. A. (1993) Building a learning organization. *Harvard Business Review*, 71(4): 78–91.

Geringer, J. M. and Hebert, L. (1989) The importance of control in international joint ventures. *Journal of International Business Studies*, (Summer): 235–54.

Ghosh, M. and John, G. (1999) Governance value analysis and marketing strategy. *Journal of Marketing*, 63: 131–45.

Ghoshal, S. and Moran, P. (1996) Bad for practice: a critique of transaction cost theory. *Academy of Management Review*, 21(1): 13–47.

Gomes-Casseres, B. (1989) Ownership structures of foreign subsidiaries: theory and evidence. *Journal of Economic Behavior and Organization*, 11: 1–25.

Granovetter, M. (1985) Economic action and social structure: a theory of embeddedness. *American Journal of Sociology*, 91: 481–510.

Gulati, R. (1995a) Does familiarity breed trust? The implications of repeated ties for contractual choice in alliances. *The Academy of Management Journal*, 38(1): 85–112.

Gulati, R. (1995b) Social structure and alliance formation patterns: a longitudinal analysis. *Administrative Science Quarterly*, 40: 619–52.

Hamel, G. (1991) Competition for competence and inter-partner learning within international strategic alliances. *Strategic Management Journal*, 12(Summer special issue): 83–103.

Hamel, G., Doz, Y. and Prahalad, C. K. (1989) Collaborate with your competitors – and win. *Harvard Business Review*, (January–February): 133–39.

Harrigan, K. (1986) *Managing for Joint Venture Success*, Lexington, MA: Lexington Books.

Hennart, J.-F. (1991) The transaction costs theory of the multinational enterprise. In C. Pitelis and R. Sugden (eds) *The Nature of the Transnational Firm*. London: Routledge, pp. 81–116.

Huber, G. P. (1991) Organizational learning: the contributing processes and literatures. *Organization Science*, 2 (1): 88–115.

Inkpen, A. C. (1996) Creating knowledge through collaboration. *California Management Review*, 39(1): 123–40.

Inkpen, A. C. and Dinur, A. (1998) Knowledge management processes and international joint ventures. *Organization Science*, 9(4): 454–68.

Kale, P., Singh, H. and Perlmutter, H. (2000) Learning and protection of proprietary assets in strategic alliances: building relational capital. *Strategic Management Journal*, 21(3): 217–38.

Khanna, T., Gulati, R. and Nohria, N. (1998) The dynamics of learning alliances: competition, cooperation and relative scope. *Strategic Management Journal*, 19(3): 193–210.

Killing, J. P. (1983) *Strategies for Joint Venture Success*, London, UK: Croom Helm.

Kogut, B. (1988) Joint ventures: theoretical and empirical perspectives. *Strategic Management Journal*, 9: 319–32.

Kogut, B. (1989) The stability of joint ventures: reciprocity and competitive rivalry. *The Journal of Industrial Economics*, 38(2): 183–98.

Kogut, B. (1991) Joint ventures and the option to expand and acquire. *Management Science*, 37(1): 19–33.

Kogut, B. and Zander, U. (1992) Knowledge of the firm, combinative capabilities and the replication of technology. *Organization Science*, 3: 383–97.

Kogut, B. and Zander, U. (1996) What firms do? Coordination, identity and learning. *Organization Science*, 7(5): 502–18.

Lepper, M. R. and Greene, D. (1975) Turning play into work: Effects of adult surveillance and extrinsic rewards on children's intrinsic motivation. *Journal of Personality and Social Psychology*, 31: 479–86.

Liebeskind, J. P. (1996) Knowledge, strategy and the theory of the firm. *Strategic Management Journal*, 17(2): 93–108.

Llewellyn, K. (1931) What price contract? An essay in perspective. *Yale Law Journal*, 40: 701–51.

Lyles, M. A. (1988) Learning among joint venture sophisticated firms. *Management International Review*, 28: 85–98.

Macneil, I. R. (1978) Contracts: adjustments of long-term economic relations under classical, neoclassical and relational contract law. *Northwestern University Law Review*, 72: 854–906.

Madhok, A. (1996) The organization of economic activity: transaction costs, firm capabilities, and the nature of governance. *Organization Science*, 7(5): 577–90.

Madhok, A. (1998) Transaction costs, firm resources and interfirm collaboration. Paper presented at the Academy of Management meetings, San Diego, CA, August 9–11.

Mahoney, J. T. and Pandian, J. R. (1992) The resource-based view within the conversation of strategic management. *Strategic Management Journal*, 13: 363–80.

Makhija, M. V. and Ganesh, U. (1997) The relationship between control and partner learning in learning-related joint ventures. *Organization Science*, 8(5): 508–27.

Mody, A. (1993) Learning through alliances. *Journal of Economic Behavior & Organization*, 20(2): 151–70.

Moran, P. and Ghoshal, S. (1996) Theories of economic organization: the case for realism and balance. *Academy of Management Review*, 21(1): 58–72.

Mowery, D. C., Oxley, J. E. and Silverman, B. S. (1996) Learning in interfirm alliances. *Strategic Management Journal*, 17(Winter special issue): 77–91.

Mowery, D. C., Oxley, J. E. and Silverman, B. S. (1998) Technological overlap and interfirm cooperation: implications for the resource-based view of the firm. *Research Policy*, 27: 507–23.

Mowery, D. C., Oxley, J. E. and Silverman, B. S. (2001) The two faces of partner-specific absorptive capacity: learning and co-specialization in strategic alliances, Harvard Business School Working Paper 01-064.

Nickerson, J. A., Hamilton, B. H. and Wada, T. (2001) Market position, resource profile and governance: linking Porter and Williamson in the context of international courier and small package services in Japan. *Strategic Management Journal*, 22(3): 251–74.

Olk, P. and Young, C. (1997) Why members stay in or leave an R&D consortium: performance and conditions of membership as determinants of continuity. *Strategic Management Journal*, 18(11): 855–77.

Oxley, J. E. (1997) Appropriability hazards and governance in strategic alliances: a transaction cost approach. *Journal of Law, Economics and Organization*, 13(2): 387–409.

Oxley, J. E. (1999) *Governance of International Strategic Alliances: Technology and Transaction Costs*. Amsterdam: Harwood Academic Publishers.

Penrose, E. T. (1959) *The Theory of the Growth of the Firm*. New York, NY: Wiley.

Pisano, G. (1989) Using equity participation to support exchange: evidence from the biotechnology industry. *Journal of Law, Economics and Organization*, 5(1): 109–26.

Pisano, G., Russo, M. and Teece, D. J. (1988) Joint ventures and collaborative arrangements in the telecommunications equipment industry. In D. Mowery (ed.) *International Collaborative Ventures in US Manufacturing*. Cambridge, MA: Ballinger, pp. 23–70.

Poppo, L. and Zenger, T. (1998) Testing alternative theories of the firm: transaction cost, knowledge-based and measurement explanations for make-or buy decisions in information services. *Strategic Management Journal*, 19(9): 853–78.

Poppo, L. and Zenger, T. (2002) Do formal contracts and relational governance function as substitutes or complements? *Strategic Management Journal*, 23(8): 707–25.

Powell, W. W. (1990) Neither market nor hierarchy: network forms of organization. In B. Staw and L. Cummings (eds) *Research in Organizational Behavior*, 12: 295–336.
Prahalad, C. K. and Hamel, G. (1990) The core competence of the corporation. *Harvard Business Review*, 68(3): 79–91.
Riordan, M. H. and Williamson, O. E. (1985) Asset specificity and economic organization. *International Journal of Industrial Organization*, 3: 365–78.
Sampson, R. (2003) The cost of misaligned governance in R&D alliances. *Journal of Law Economics and Organization* (Forthcoming).
Shane, S. (1994) The effect of national culture on the choice between licensing and direct foreign investment. *Strategic Management Journal*, 15: 627–42.
Simonin, B. L. (1997) The importance of collaborative know-how: an empirical test of the learning organization. *Academy of Management Journal*, 40(5): 1150–74.
Sobrero, M. and Roberts, E. B. (2001) The trade-off between efficiency and learning in inter-organizational relationships for product development. *Management Science*, 47(4): 493–511.
Strickland, L. H. (1958) Surveillance and trust. *Journal of Personality*, 26: 200–15.
Teece, D. J. (1977) Technology transfer by multinational firms: the resource costs of transferring technological know-how. *The Economic Journal*, (June): 242–61.
Teece, D. J. (1981) The market for know-how and the efficient international transfer of technology. *Annals of the American Academy of Political and Social Science*, (November): 81–96.
Teece, D. J. (1986) Profiting from technological innovation: implications for integration, collaboration, licensing and public policy. *Research Policy*, 15: 285–305.
Teece, D. J. and Pisano, G. (1994) The dynamic capabilities of firms: an introduction. *Industrial and Corporate Change*, 3: 537–56.
Teece, D. J., Pisano, G. and Shuen, A. (1997) Dynamic capabilities and strategic management. *Strategic Management Journal*, 18(7): 509–33.
Wernerfelt, B. (1984) A resource-based view of the firm. *Strategic Management Journal*, 5(2): 171–80.
Williamson, O. E. (1975) *Markets and Hierarchies*. New York: Free Press.
Williamson, O. E. (1983) Credible commitments: using hostages to support exchange. *American Economic Review*, 73: 519–40.
Williamson, O. E. (1985) *The Economic Institutions of Capitalism*. New York, NY: Free Press.
Williamson, O. E. (1991) Comparative economic organization – the analysis of discrete structural alternatives. *Administrative Science Quarterly*, 36(4): 269–96.
Williamson, O. E. (1996a) *The Mechanisms of Governance*. Oxford: Oxford University Press.
Williamson, O. E. (1996b) Economic organization: the case for candor. *Academy of Management Review*, 21(1): 48–57.
Zajac, E. J. and Olsen, C. P. (1993) From transaction cost to transaction value analysis: Implications for the study of interorganizational strategies. *Journal of Management Studies*, 30(1): 131–45.

7

Relationship Dynamics: Developing Business Relationships and Creating Value

Ulf Andersson and Benjamin Ståhl

Introduction

In 1997, a world-leading tooling supplier lost a large customer. Virtually overnight, and without consultation, the customer announced that it would be moving all of its tooling supplies to an integrator. This would distance the supplier from the customer, and indirectly decrease volumes and margins. This dramatic event prompted the supplier to come up with a new strategy to bind customers closer to them, a strategy that came to change the way that they did business and which was highly successful.

The motivation for customers to use integrators – 'one stop shops' – was mainly to regain control of their tooling supplies in terms of inventory control, tool standardization and more efficient transaction handling (orders and invoicing). The supplier would have to offer similar services to stem the flow of business going through integrators. They decided to develop a 'preferred supplier' marketing concept that offered such benefits to customers. Furthermore, they offered technical help to customers with the aim of optimizing their operations, in return for conversion to their tools. They were in a better position to undertake such optimization than their customers since had both a wider knowledge about tooling applications and the existing products.

However, this offer entailed much more resource commitment in terms of human resources from the supplier. Furthermore, the offer had to be trustworthy in the sense of being able to demonstrate and validate cost savings made. To accomplish these two ends, a methodology for optimizing customer operations was developed. The method – PCA for

Productivity Cost Analysis – was a relatively simple piece of software that application engineers could use in the field, enabling them to benchmark existing operations and suggest improvements relatively quickly. The reports generated also showed customers bottom-line improvement in terms of costs and productivity. For inventory control and cost savings, other methods were devised, such as consignment stock. Especially important was an alliance with a supplier of automatic tool dispensers – 'vending machines' for tools, providing continuous and exact reporting of inventory and tool usage.

The preferred supplier concept has been quite successful. Sales volume in agreement accounts have risen 83 per cent in a recessive market, as competitor products have been replaced. For customers, the expected annual cost savings of 20 per cent or more, in terms of inventory control and productivity increase due to optimization, have been met.

This case exposes how strong and close business relationship can create value for both parties to an exchange. This is not a new phenomenon, and neither has it gone unnoticed in theory. Despite the importance of the topic and the existence of multiple frameworks concerning *why* firms engage in business relationships, there is a lack of research concerning *how* relationships are developed (Ring and Van de Ven, 1994). Most research is purely conceptual (ibid.), or induced from only a few cases. The same is true for research on how capabilities emerge (Verona, 1999). In both cases, very little inter-industrial, cross-sectional empirical research has been conducted. There thus exists a clear need to empirically investigate the dynamics of interorganizational relationships and capabilities (Ring and Van de Ven, 1994). This paper concerns how business relationships that deliver value to both parties are developed.

The next section provides a theoretical discussion of business relationships. The following section uses theory and the case study to develop hypotheses and a structural model of interorganizational relationship development. We then test the model on data from 277 customer relationships, followed by a discussion of the implications for theory and management.

Business relationships

Economic organization concerns the combination, allocation and exchange of resources. In its simplest abstraction, economic organization occurs through markets or hierarchical arrangements, with their corollaries price and authority (Coase, 1937). However, this dichotomy is quite unsatisfactory for theorists and practitioners of business. Early challenges include Thompson (1967), and the so-called interaction approach (e.g. Håkansson, 1993). In common these and other perspectives argue that resources are heterogeneous and interdependent. The value of a resource depends on its combination with other resources and the activities surrounding it. Heterogeneity of resources entails a high degree of specialization, and specialization in turn gives rise to further heterogeneity (Håkansson, 1993). Thus, knowledge about resources is also heterogeneous and distributed. While prices constitute one kind of information concerning a resource or a product, more or less idiosyncratic knowledge about how to produce it and how to use it (Dahlqvist, 1998) usually *interact* for resource development to occur.

Much of such interaction has been shown to take place in business relationships, which means that a common mode of economic organization includes dimensions both of market and hierarchy. This means that firms operate in an interdependent system, or network. As Dyer and Singh (1998) state, the (dis)advantages of an individual firm are often linked to the (dis)advantages of the network of relationships in which the firm is embedded.

Dyer and Singh (1998) define relational rents as consisting of supernormal profit jointly generated in an exchange relationship that cannot be generated by either firm in isolation and can only be created through the joint idiosyncratic contributions of the specific alliance partners. The use of the term 'alliances' reflects one stream of research that is especially interested in the emergence of horizontal cooperative interorganizational relationships, such as strategic alliances between 'competitors'. Another stream has instead focused on vertical relationships between suppliers and customers (Håkansson and Snehota, 1990). The definition offered by Dyer and Singh (1998) is, however, equally applicable in these cases.

They further posit that relational rents are a function of the following:

> (1) investments in relation-specific assets; (2) substantial knowledge exchange, including the exchange of knowledge that results in joint learning; (3) the combining of complementary, but scarce, resources or capabilities (typically through multiple functional interfaces), which results in the joint creation of unique new products, services, or technologies; and (4) lower transaction costs than competitor alliances, owing to more effective governance mechanisms. (Dyer and Singh, 1998)

While these are structural characteristics, the rationale for business relationships, we now turn to another approach. This approach focuses on emergent collective learning and routine-creation accompanying close and frequent exchange between firms (Håkansson, 1993).

While price is always an important dimension of exchange even in long-term, stable business relationships, a more important feature is interaction aiming to increase the overall efficiency, through adaptation and development of the resources and activities that the respective actors are involved in. Studies have shown conclusively that interorganizational relationships are arrangements through which the business process can be effectively managed and above normal rents created and sustained (Levinthal and Fichman, 1988; Zajac and Olsen, 1993; Anderson *et al.*, 1994; Dyer and Singh, 1998). Exchange effectiveness is improved through adaptation between business actors (Hallén *et al.*, 1991), investments in relationship-specific activities for higher efficiency (Dyer and Singh, 1998), establishment of cross-functional communication between business actors (Olson *et al.*, 1995; Ragatz *et al.*, 1997), and enhanced understanding of the production system in which exchanged products are to be used (Mattson, 1978; Dahlqvist, 1998).

Stability in terms of business counterparts does not indicate a lack of dynamism, however. Most business actors find themselves in a situation where continuous development in terms of product development, innovation, changing customer requirements, changing supplier offering, new markets, etc. Thus, the content of business relationships is often very dynamic in terms of development of the resources and activities involved. Business relationships are thus vehicles for interaction, enabling actors to develop products and production processes (Lundvall, 1985; von Hippel, 1988) and exchange information about business opportunities (Ottum and Moore, 1997).

But even in business-to-business exchange, not all products or services are associated with deep interaction, strong ties and long-term stability. Sometimes, as in the case discussed in this paper, this is due to the consumable and relatively standardized nature of the exchanged good, and the presence of several competitors with similar offerings, i.e., the price of each exchanged product is low (almost negligible) relative to total costs. The wide range of specifications allows suppliers to make frequent calls offering the latest versions and, more importantly, the latest discounts. Thus, business exchange is of an almost arms-length character, with price as a major decision-making factor.

However, the resulting situation is sub optimal for both customers and suppliers. By focusing only on the price dimension, customers have little control over their supply base and incur high indirect costs relating to inventory control, logistics and fragmented production processes. Similarly, suppliers are caught in a discounting race that erodes their margins and makes it difficult to recover development costs. Higher levels of interaction can then by used to create closer ties, higher stability and superior performance.

Specifically, this chapter addresses areas that have not been exhaustively studied. The dominant line of research on relationships has focused on such motivational factors as trust and commitment (Anderson and Narus, 1994; Morgan and Hunt, 1994; Ring and Van de Ven, 1992, 1994), and less on the underlying value created in business relationships. Furthermore, while we know much about why relationships exist, we know much less about how they emerge (Ring and Van de Ven, 1994). In sum, recognizing the interdependencies inherent in systems of production and knowledge, mediated through the existence of business relationships, poses new questions concerning business behaviour in general and strategy in particular (Håkansson and Snehota, 1990; Dyer and Singh, 1998). The question we focus on in this paper is this: If performance can be related to a firm's ability to develop relationships, how are capabilities unreachable for any firm in isolation mobilized?

A model of relationship development

In the case referred to in the beginning of this chapter,[1] the gains from the business relationship were cost savings and productivity increases for the customer and higher sales volumes for the supplier. The driver for these outcomes were changes made in the production system of the customer, in turn dependent on changes in the supplier's sales and marketing methods. The outcomes do not relate specifically to product development at the customer, but rather efficiency-enhancing knowledge creation. Application engineers from the supplier spend most of their time with the customer, carrying out experimentation, tuning, hardware and software installation, aiming to optimize machine operations and stock control.

A basic driver of the relationship development is the frequent interaction between the customer and supplier, establishing common working methods, norms and understanding of the counterpart's needs and capabilities. In an interorganizational setting, interaction and socialization is in itself a part of negotiations. The French social psychologist Moscovici introduced the concept of 'social representations' (Moscovici, 1976, 1988), conceived of as socially elaborated and shared knowledge forms that have practical aims and concur with the construction of a reality common to a social group (Moscovici, 1988). Social representations are information carriers, and they are important for individuals' understanding and communicating.

In a business setting social representations could be thought of as 'shared business perceptions' emerging as customers and suppliers start to understand how they could benefit from exchange with each other. When actors develop new businesses they need to communicate knowledge for which no accurate concepts have been developed, and to which no easily understood perceptions correspond. As noted by Kirzner (1973), knowledge about new business opportunities is knowledge about reality under construction, which implies that it is knowledge about something that business actors have hitherto been ignorant – it is a process of discovery. Interaction may create shared business perceptions that allow actors to carry communication, interaction, and exchange a step forward. Knowledge associated with social representations is knowledge whose objective is to create reality (Moscovici, 1988: 229).

In the case of the preferred supplier agreements, the customers put a high value on the supplier's application engineers' presence in their companies. The supplier representatives have offices inside the customers' production facilities, spend much time on the shop floor and are always on call. Such interaction gave the supplier a very good understanding of the customers' processes and needs, and was fundamental to building trust with the customers' machine operators. We term this *interaction intensity*.

> **Hypothesis 1**: High interaction intensity in a business relationship will have a positive effect on relationship development.

However, in the case of the preferred supplier relationships described in the case, interaction *alone* cannot explain the strength of the relationship. Social interaction and communication were mostly connected to specific activities taking place, such as benchmarking an operation, changing tooling or going over the weekly results. In other words, shared business perceptions develop as business actors impose some organization on their material and social environment, validate their knowledge of it, and confirm the rules they have adopted for dealing with it. Interaction was purposeful and pragmatically oriented.

When two business actors meet, the only way in which they can communicate and adopt mutually satisfactory behaviours is by filtering a few predominant features out of the mass of inchoate impressions: intentions, feelings, capacities, and so on. Material objects give rise to the same operation, involving comparison, classification, and selection of the transmitted stimuli. In the case of the preferred supplier relationships, the counterparts engaged in trial and error, benchmarking, tool conversions, hardware installations, etc. Furthermore, these processes were meticulously documented, with the help of software, and the reports generated were scrutinized and analyzed together. This is similar to the process that has been labelled externalization (Nonaka and Takeuchi, 1986), whereby tacit knowledge is transformed into explicit. Externalization essentially depends on codification, whereby tacit knowledge can be disseminated more easily and thereby accessible on another level and to an increased number of people.

As such, it is possible to combine knowledge and create new solutions. In an inter-organizational setting, an expression of such combination is thus adaptations in products or production technology. By such alterations, the capabilities of the business partners become better aligned and boost efficiency, while at the same time promoting the idiosyncratic nature of the relationship.

Regarding the externalization, we can conclude that not only is it important that engaged business actors are allowed to express their viewpoints: they also need to be involved in *task-driven co-action*. Findings on capability development have pointed to the importance of continuous experimentation and prototyping (Leonard-Barton, 1995). This helps to avoid destructive conflict and establish fair behaviour where reciprocity and interdependence is accepted, and where actors not only try to influence others but are also influenced themselves. Task-driven co-action should be understood as continuous externalization of the ideas discussed in the business relationship. Task-driven co-action brings reciprocity and interdependence between business actors to the surface and limits the risk of intellectual conflict between differing viewpoints – conflict that might result in action paralysis. In other words, even if it is important to establish genuine dialogue, to listen to each other, task-driven co-action is indispensable for the development of shared business perceptions. Task-driven co-action fosters understanding of interdependencies in the business exchange, and in the end relationship development is affected.

> **Hypothesis 2**: Task-driven co-action in a business relationship will have a positive effect on relationship development.

Task-driven co-action has a direct effect on relationship development, acting as a catalyst for fair behaviour and appreciation of interdependencies. As ideas are put into practice, an incremental trouble-shooting process is set in motion (Malmberg *et al.*, 1996) resulting in a need to discuss consequences not previously considered. Specific unpredicted interdependencies and outcomes are brought to the surface and fuel interaction. The use of prototyping in product development is an example where this effect is deliberately activated (Tabrizi and Walleigh, 1997). In the case of the preferred supplier agreements, the documents and reports generated by software and application engineers' benchmarking and experimentation activities were natural and recurring foundations for discussions and decisions. Thus, task-driven co-action will have an impact on interaction intensity.

> **Hypothesis 3**: Task-driven co-action in a business relationship will have a positive effect on interaction intensity.

In this view interaction and communication are not confined to verbal discussions, but it is also important that the actors have the possibility to work together and continuously materialize their ideas. Since relationship development concerns ill-understood future action, satisfying communication cannot be secured only through eloquent talk.

The discussed hypotheses are summarized in Figure 7.1, where the numbers correspond to the presented hypotheses.

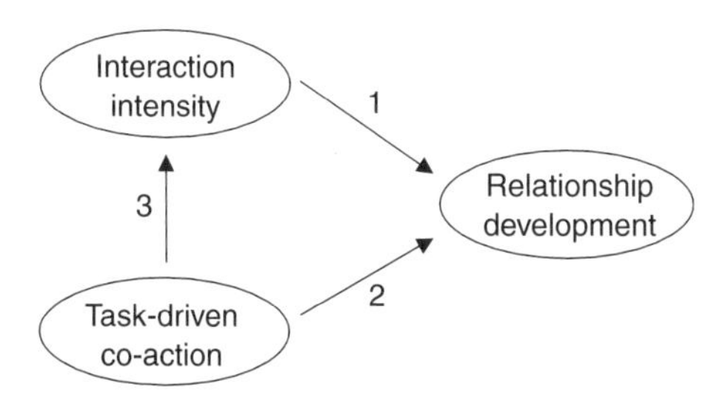

Figure 7.1 A structural model of the relations between interaction intensity, task-driven co-action and relationship development

Data and method

In this section the variables needed for testing the above hypotheses are operationalized. The hypotheses are then tested in a LISREL model.

The sample used is a sub-sample of a larger study on subsidiaries of Swedish multinational companies. In the original study subsidiary networks and their importance for and influence on the subsidiaries' development in a number of different aspects were investigated (Pahlberg, 1996; Andersson, 1997; Thilenius, 1998). In this particular study we have chosen to concentrate on the subsidiaries' customer relationships. The suppliers are situated in diverse industries, such as gas applications, hard material tools, industrial equipment, managerial training, petrochemicals, power distribution, pulp and paper, software and telecommunication equipment. The size of the suppliers varies between 50 and more than 5000 employees. The sample used consists of 277 customer relationships with 97 suppliers located in Europe and North America. To increase validity and reliability each of the 97 suppliers has been personally visited and interviewed using a standardized questionnaire, each interview taking approximately two hours. By conducting personal interviews the researcher has the possibility to discuss and explain problems with concepts and interpretations in the questionnaire with the respondent. This possibility clearly improves the reliability of the answers collected in comparison with for example a mail survey.

The standardized questionnaire used five-point Likert-type scales, ranging from 'not at all' to 'very much' or from 'not at all' to 'very high'. Missing values in the analysis were accounted for by listwise deletion (Jöreskog and Sörbom, 1993).

Construct analysis

The hypothesized model shown in Figure 7.1 was empirically tested in a LISREL model. LISREL models are judged by three different kinds of validity, nomological, discriminant, and convergent. Nomological validity concerns the overall fit between the proposed model and the data. Nomological validity is assessed by chi-square (χ^2) together with degree of freedom measures, and a probability estimate, (p-value). The significance of the model is estimated by the p-value, which should exceed 0.05 for significance on the 95 per cent level. The distance between the proposed model and the data is measured by the χ^2 measurement together with the degree of freedom measurement (Jöreskog and Sörbom, 1993).

Convergent validity measures the homogeneity of the constructs used in the model and discriminant validity the degree to which the constructs are separated from each other. Discriminant and convergent validity are determined by studying the R^2-values, measuring the strength of a linear relationship, and t-values, a significance test, of each relationship in the model (ibid.) The results of the tests of the constructs convergent validity are shown in Table 7.1 (p. 140). To assess discriminant validity a model with no causal relations between the constructs, a so-called measurement model, was constructed. The latent constructs in the model are discriminantly valid as key statistical estimates show that no pair of constructs is unidimensional. The following section describes how the constructs were operationalized.

Relationship development

It is important to keep in mind that in this paper relationship development is not equated with product development. The important aspect of relationship development concerns a business actor's improving capabilities to demonstrate to its customer that it is able to deliver value. Whether this is achieved by improving material features in products supplied or by enhancing the efficiency of the customer's production process is an open question. Relationship development means describing and modifying the functions of a product in ways that deliver more value to the customer. Relationship development, as conceived of here, is concomitant with knowledge creation in an interorganizational setting. Development of new businesses, just as in the case of development of new ways of thinking and acting in society in general, demands serious reconsideration of what to do and how to do it. Relationship development is produced by actors devoted to the exploration of possibilities associated with controlled resources.

When relationship development is operationalized it is important to capture a phenomenon that accounts for appreciation of knowledge about counterparts, ability to modify behaviour, and change in exchange effectiveness. Three indicators have been used, namely supply of market information from the customer, adaptation in business conduct in relation to the customer, and the customer's importance for sales volume. Since a business phenomenon is being discussed in the chapter it is important to capture the economic aspect of relationship development. Relationship development does not just refer to change in attitude, but also to change in performance. The customer's importance for sales volume has been used to capture this aspect of relationship development. As has been described above, innovative behaviour is the result of actors being able to listen to and evaluate information provided by other actors, and then to re-evaluate their own knowledge and opinions. Supply of market information from the customer has been used as an indicator for the information aspect of relationship development. When suppliers understand possibilities and opportunities

Table 7.1 The constructs and their indicators

Indicator	*Factor loading*	t-*value*	R^2-*value*
Interaction intensity			
Number of times people from the supplier are involved in direct contact with people from this customer during a year (TIMES)	0.39	4.29	0.15
Number of different knowledge areas involved in direct contact with people from this customer (KNWA)	0.58	4.48	0.34
Task-driven co-action			
To what extent has the relationship caused adaptation for the supplier concerning:			
–product technology (APT)?	0.94	15.28	0.88
–production technology (APN)?	0.72	11.66	0.51
Relationship development			
To what extent is this customer important to the supplier concerning market information (MINF)?	0.50	6.44	0.25
To what extent has the relationship with this customer caused adaptation for the supplier concerning business conduct (ABC)?	0.85	5.86	0.73
To what extent is this customer important to the supplier concerning sales volume (S_VOL)?	0.63	7.25	0.40

Note: Abbreviations in parenthesis are indicator names used in Figure 7.2.

demonstrated by other actors they need to be able to respond by changing behaviour. Business actors need to change the way they do business and carry out business opportunities. Adaptation in business conduct has been used as an indicator for the behavioural aspect of relationship development. Taken together, 'relationship development' refers to increased business performance through simultaneous management of sales, information processing, and business conduct.

Key statistics for the three indicators used for the latent construct 'relationship development' are shown in Table 7.1. The indicators are valid representations of the construct 'relationship development'. All key statistical measures are good. The *t*-values are above 5.86, factor loadings are above 0.50, and the R^2-values are above 0.25. The *t*-values and R^2-values indicate good convergent validity.

Interaction intensity

The development of new businesses is dependent on interaction that fosters genuine dialogue where diverging viewpoints, opinions, judgements, and ideas are allowed expression and are cross-examined. From the previous discussion it follows that establishment of genuine dialogue is dependent on actors' ability to demonstrate behaviour, where presented ideas as well as the presenting actor, are experienced as trustworthy and sensible – in other words, that they can deliver value to the customer. Since relationship development involves creating what did not previously exist, it is important that involved actors have the opportunity to consistently reformulate and rephrase their views of how to develop new businesses.

In business relationships the prime focus of interaction and associated discussions is enhanced performance. In those relationships where we can talk about business development, enhanced performance is managed through establishment of new ways of using existing resources. Interaction revolves around the possibility of developing something that could improve business performance.

When interaction intensity is operationalized it is important to capture a phenomenon that covers frequency and quality. Two indicators have been used for the construct 'interaction intensity'. First, the number of times people representing the supplier have been involved in direct contact with people representing the customer, and secondly, the number of different knowledge areas involved in direct contact with this customer. The possibility of demonstrating consistency and rephrasing and reformulating presented business opportunities is, to a large extent, a matter of frequent meetings. The number of direct contacts has therefore been used as an indicator for the consistency aspect of interaction intensity. If optional resources that could otherwise have been used in other activities were supplied it was conjectured that business actors demonstrated investment. In a business relationship the number of knowledgeable and relevant people that are engaged by a business actor constitute a straightforward indicator for the willingness to invest in the relationship. Moreover, following the idea of Nonaka regarding the value of 'redundant' knowledge and information as conducive to knowledge creation, issues concerning many different aspects of the work may be brought to the surface (Nonaka and Takeuchi, 1986). Examples include the possibility of developing a prototype, the possibility of producing the product, the possibility of marketing it, and the possibility of using an existing product in new ways. Therefore, it is important to staff the relationship with individuals representing different knowledge areas in the business firm. The number of involved knowledge areas was used as an additional indicator for the investment aspect of interaction intensity.

The above two indicators are not used as indicators for a construct that describes intense, complex communication as effective *per se* for the establishment of shared business perceptions and the development of new businesses. Rather, the construct suggests that it is important for business actors to be able to allocate relevant knowledge as different concerns surface during the interaction. The development of shared business perceptions is a fragile operation, and it may be obstructed if uncertainties about the desirability and intelligibility of discussed business opportunities cannot be properly addressed.

The indicators of interaction intensity seem to be valid representations of a common construct. Key statistics for the two indicators used for the latent variable 'interaction intensity' are shown in Table 7.1. All key statistical measures are good except the strength of the linear relationship for the indicator 'times' (R^2-value 0.15). Due to the conceptual importance of this indicator it was decided to retain times as an indicator of the latent variable 'interaction intensity'. The *t*-values are 4.29 or higher, factor loadings are above 0.39, the R^2-value is 0.34 for the indicator knowledge areas and, as mentioned above, 0.15 for the indicator times. The implication of the statistical results is that the construct 'interaction intensity' is validly represented by the indicators chosen, but that the indicator 'knowledge areas' is more central to the construct than the indicator 'times' is, a finding further emphasizing the importance of redundancy and combinative capabilities for knowledge creation.

Task-driven co-action

The most effective way to settle a discussion concerning probable consequences of proposed ideas is to try to materialize them. Here, task-driven co-action is viewed as the activities that aim at realization of proposed business ideas into actual business opportunities. During these attempts actors become aware of interdependence between different areas of the business exchange (Tabrizi and Walleigh, 1997). Effective knowledge creation entails that knowledge is codified so that it may be disseminated, tested and combined with other solutions (Nonaka and Takeuchi, 1986). As such, task-driven co-action constitutes the practical aspect of interaction. Task-driven co-action is not seen here as the completion of ideas discussed in a business relationship, but rather as incremental testing and prototyping along the way.

When task-driven co-action is operationalized it is important to capture a phenomenon whereby proposed ideas are actually realized. Two indicators have been used for the construct 'task-driven co-action', namely adaptations made in product technology and adaptations made in production technology. Adaptations made in product technology and production technology constitute materializations of discussed ideas and result in unmistakable consequences that could not be discredited by intellectual arguments.

Key statistics for the two indicators used for the latent variable 'task-driven co-action' are shown in Table 7.1. The indicators for 'task-driven co-action' seem to be valid representations of a common construct. All key statistical measures are good. The *t*-values are above 11.66, factor loadings are 0.72 or higher, and the R^2-values are above 0.51. The *t*-values and R^2-values indicate good convergent validity.

Results

Exchange effectiveness is dependent on the ability of the actors to negotiate viable exchange agreements. The process of negotiation was conceived of as a knowledge creation process labelled 'relationship development'. In this paper relationship development was distinguished from product development in the sense that development of new arguments as well as the development of new production and exchange methods contributed to relationship development. It was hypothesized that the existence of intense interaction and task-driven co-action in a business relationship directly affected relationship development. It was further hypothesized that the existence of task-driven co-action indirectly affected relationship development via interaction intensity. The rationale for this latter hypothesis was that as business actors try to realize business ideas discussed at the verbal level unforeseen consequences may surface, which have to be further discussed. These hypotheses were summarized in Figure 7.1.

The result of the LISREL analysis displayed in Figure 7.2 represents a model with a *chi*-square value of 13.77 at 11 degrees of freedom and a probability value of 0.25. The probability value above 0.05, and the *chi*-square and degrees of freedom measures propose that the model fits the data well. Further support for this is found in the fact that all causal relationships between latent variables have *t*-values above 2.38 and R^2-values above 0.21.

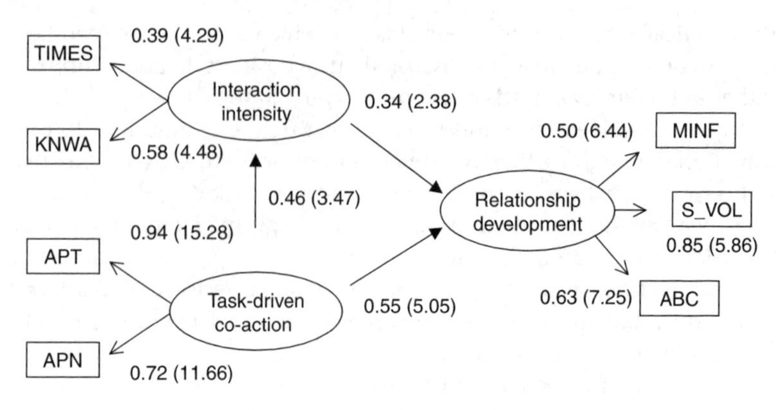

Figure 7.2 A LISREL model of relationship development

Note: Figures are factor loadings, with t-values in parenthesis. Chi-square is 13.77, with 11 degrees of freedom, at a probability of 0.25.

From the LISREL analysis we see that 'interaction intensity' as well as 'task-driven co-action' has a strong impact on 'relationship development'. The analysis demonstrates that 'task-driven co-action' (with a coefficient of 0.55 and a *t*-value of 5.05) has a stronger impact than 'interaction intensity' (with a coefficient of 0.34 and a *t*-value of 2.38). That could be seen as an indication that important business knowledge is difficult to manage and transfer in any straightforward manner, but is rather mediated through activities carried out between business actors. Further, the analysis shows that 'task-driven co-action' impacts on 'relationship development' via 'interaction intensity' (with a coefficient of 0.46 and a *t*-value of 3.47), a result that is consistent with the idea that externalization is an important aspect of meaningful interaction to take place.

Discussion

Analysis of the structural model demonstrates that relationship development is dependent not only on interaction but also real co-action, i.e. the materializing of ideas. The analysis further indicates that knowledge may be difficult to transfer between business actors, but could be mediated through co-action between involved actors. The business relationship could thus be conceived of as an organizational system that effectively mediates business knowledge and promotes business exchange.

The analysis conducted in this paper is based upon a sample of business relationships that are considered to be important by the actors involved.

Even though the analysis clearly demonstrates that close interaction between business actors constitutes an important aspect of effective business exchange it does suggest a need for future examination of the mutuality of business exchange. The current analysis is based on the supplier's behaviour towards the customers, and further studies that investigate the simultaneous behaviour of the supplier as well as the customer will shed further light on the process by which new business opportunities are continuously developed in our economy. Another aspect of relationship development through counterpart interaction that should be addressed concerns the impact of cognitive distance between involved actors. As indicated by studies made of innovation and business development (Mieszkowski, 1998; Moscovici, 1976; Moscovici and Faucheux, 1972) the existence of internal conflict in a group creates a predisposition towards and potential for, change in cognition and action. This means that business development through counterpart interaction encompasses two aspects. First, the existence of diverging viewpoints and secondly, the existence of interaction styles that enables involved actors to appreciate and change these viewpoints. Empirical studies that manage to investigate both of these aspects simultaneously will indeed push our understanding of business behaviour a step forward.

While not denying that internal organizational capabilities indeed matter, as they may further leverage knowledge creation, e.g. the pace of product development (Verona, 1999), we would argue that a capabilities perspective forces us to consider interorganizational capabilities *per se*. In other words, business relationships can be a conduit to create new capabilities on the interorganizational level as well as enhancing the internal capabilities of the partners. Similarly, notions of 'absorbing' or 'sourcing' knowledge may in some cases be misleading. Concerning interorganizational relationships and capabilities, they require cooperation and commitment from both partners, and together they create value, which would be unattainable and unsustainable for either part alone.

According to Srivastava *et al.* (1998) market-based assets such as customer relationships, and market knowledge generated by these relationships, are of great importance for a firm's ability to develop and market products. Hereby market-based assets, these authors argue, impact on the firm's ability to accelerate and enhance cash flows, and in the end generation of shareholder value is influenced. Customer relationships do not automatically convert into valuable assets; only when the customer relationship enables the firm to take more effective action in the market place could it be said to constitute an asset.

Whether a relationship should in fact qualify as a market-based asset depends, to a large extent, on the firm's ability simultaneously to make use of knowledge about customer preferences and knowledge about production capabilities. The research presented in this study demonstrates that this possibility is highly dependent on the way the interaction with the customer is managed. Frequently, knowledge relevant for development of marketing capabilities turns out to be difficult to decontextualize and transfer between actors. The existence of close and continuous interaction with the customer where intense interaction and task-driven co-action could be established is imperative to the firm's ability to effectively capitalize on the customer relationship. Management of external relationships becomes in this view an important issue for the firm's strategic work, and properly managed they will yield to a firm's market-based assets.

For the practitioner, the insight that management of external relationships has to be incorporated into the firm's business strategy provides a first clue as how to build market-based assets out of a firm's counterpart interaction. Starting from strategic analysis of opportunities and challenges facing the firm, a focus on existent business relationships could be used to identify the relationships most likely to add value to the firm in terms of intellectual input to the firm's capacity to develop new businesses. Such an analysis could complement work dealing with the establishment of knowledge strategies that are observed in many of today's firms. The focus on existing relationships, and the value provided by those relationships, constitutes a substantial method to align knowledge strategy with overall business strategy.

The results presented in the chapter indicate that when a firm's strategic relationships are identified the next step in the endeavour to build market-based assets is to manage the relationships in terms of interaction and task-driven co-action. To the extent that resources are available, continuous contacts with people from exchange counterparts, where individuals possessing the relevant experience have the possibility of testing out ideas for new business opportunities, will have a positive impact on future performance. Proper management of relationships decreases the risk that products that are launched will fail, that important service offerings to customers will be overlooked, or that the possibility of rendering exchange activities more effective will be disregarded.

Interaction alone, however, is insufficient. The results of the present study indicate that externalization of ideas is an important aspect of developing business relationships. Concrete adaptations, documentation of tests, reports, prototyping and the like not only promote the relationship directly, but also indirectly by providing the base for further interaction. Such co-action discloses and manifests the value creation of the relationship, making further business more likely. Careful attention to routines and methods enhancing co-action is thus important for relationship management.

This chapter constitutes a first guide for identification of strategic relationships, and for establishment of relationship strategies that will increase the firm's ability to refine customer relationships into market-based assets.

Note

1. The case will be reported in depth elsewhere. It involves the development of a preferred supplier agreement at the UK subsidiary of a world-leading industrial tooling company, and the subsequent intra-organizational transfer of the concept. In total, 18 interviews in 5 countries were conducted with the company, and 3 customers were visited. The company has requested to remain anonymous for the time being.

References

Anderson, J., Håkansson, H. and Johanson, J. (1994) Dyadic business relationships within a business network context. *Journal of Marketing*, 58: 1–15.

Anderson, J. and Narus, J. (1994) A model of distributor firm and manufacturer firm working partnerships. *Journal of Marketing*, 54: 42–58.

Andersson, U. (1997) *Subsidiary Network Embeddedness: Integration, Control and Influence in the Multinational Corporation*. Uppsala: Department of Business Studies, Uppsala University.

Coase, R. H. (1937) The Nature of the Firm. *Economica*, n.s. 4(November): 386–405.

Dahlqvist, J. (1998) *Knowledge Use in Business Exchange*. Uppsala: Department of Business Studies, Uppsala University.

Dyer, J. and Singh, H. (1998) The relational view: cooperative strategy and sources of interorganizational advantage. *Academy of Management Review*, 23(4): 660–79.

Håkansson, H. (1993) Networks as a mechanism to develop resources. In P. Beije *et al.* (eds) *Networking in Dutch Industries*. Leren-Apendorn: Garant.

Håkansson, H. and Snehota, I. (1990) No business is an island. *Scandinavian Journal of Management*, 5: 187–200.

Hallén, L., Johanson, J. and Sayed-Mohamed, N. (1991) Interfirm adaptations in business relationships. *Journal of Marketing*, 55: 29–37.

Hippel, E. von (1988) *The Sources of Innovation*. New York: Oxford University Press.

Jöreskog, K. G. and Sörbom, D. (1993) *LISREL 8: Structural Equation Modeling with the SIMPLIS Command Language*. Chicago: Scientific Software International.

Kirzner, I. M. (1973) *Competition and Entrepreneurship*. Chicago: University of Chicago Press.

Leonard-Barton, D. (1995) *Wellsprings of Knowledge: Building and Sustaining the Sources of Innovation*. Boston: Harvard Business School Press.

Levinthal, D. A. and Fichman, M. (1988) Dynamics and inter-organizational attachments: auditor–client relationships. *Administrative Science Quarterly*, 33: 345–69.

Lundvall, B.-Å. (1985) *Product Innovation and User-Producer Interaction*. Aalborg: Aalborg University Press.

Malmberg, A., Sölvell, Ö. and Zander, I. (1996) Spatial clustering, local accumulation of knowledge and firm competitiveness. *Geografiska Annaler*, 78B: 85–97.

Mattson, L. G. (1978) Impact of stability in supplier–buyer relations on innovative behaviour on industrial markets. In G. Fisk, J. Arndt and K. Grönhaug (eds) *Future Direction for Marketing*. Cambridge, MA: Marketing Science Institute.

Mieszkowski, K. (1998) Opposites attract. *Fast Company*, November–December: 42–4.

Morgan, R. M. and Hunt, S. D. (1994) The commitment: trust theory of relationship marketing. *Journal of Marketing*, 58: 20–38.

Moscovici, S. (1976) *Social Influence and Social Change*. London: Academic Press.

Moscovici, S. (1988) Notes towards a description of social representations. *European Journal of Sociology*, 18: 211–50.

Moscovici, S. and Faucheux, C. (1972) Social influence, conformity bias, and the study of active minority. In L. Berkowitz (ed.) *Advances in Experimental Social Psychology*. New York: Academic Press.

Nonaka, I. and Takeuchi, H. (1986) *The Knowledge-Creating Company*. New York: Oxford University Press.

Olson, E. M., Walker, C. Jr. and Ruekert, R. W. (1995) Organizing for effective new product development: the moderating role of product innovativeness. *Journal of Marketing*, 59: 48–62.

Ottum, B. D. and Moore, W. L. (1997) The role of market information in new product success/failure. *Journal of Product Innovation Management*, 14: 258–73.

Pahlberg, C. (1996) *Subsidiary–Headquarters Relationships in International Business Networks*. Uppsala: Department of Business Studies, Uppsala University.

Ragatz, G. L., Handfield, R. B. and Scannell, T. V. (1997) Success factors for integrating suppliers into new product development. *Journal of Product Innovation Management*, 14: 190–202.

Ring, Peter S. and Van de Ven, A. (1992) Structuring Cooperative Relationships Between Organizations. *Strategic Management Journal*, 13: 483–98.

Ring, P. S. and Van de Ven, A. (1994) Developmental processes of cooperative inter-organizational relationships. *Academy of Management Review*, 19: 90–118.

Srivastava, K. R., Shervani, T. A. and Fahey, L. (1998) Market-based assets and shareholder value: a framework for analysis. *Journal of Marketing*, 62: 2–18.

Tabrizi, B. and Walleigh, R. (1997) Defining next-generation products: an inside look. *Harvard Business Review*, November–December: 116–24.

Thompson, J. D. (1967) *Organizations in Action*. New York: McGraw-Hill.

Thilenius, P. (1998) *Subsidiary Network Context in International Firms*. Uppsala: Department of Business Studies, Uppsala University.

Verona, G. (1999) A Resource-based view of product management. *Academy of Management Review*, 24(1): 132–42.

Zajac, E. J. and Olsen, C. P. (1993) From transaction cost to transaction value analysis: implications for the study of interorganizational strategies. *Journal of Management Studies*, 30: 131–45.

Part III

Governing Internal Knowledge Relations

8

Identifying Leading-Edge Market Knowledge in Multinational Corporations

Niklas Arvidsson and Julian Birkinshaw

Introduction

In his seminal article, 'The hypermodern MNC – a heterarchy?' Gunnar Hedlund (1986) argued that one of the distinctive features of the heterarchical multinational corporation (MNC) is its holographic qualities. By this he meant that knowledge about the whole is contained in each part, which is very different from the archetypal hierarchical organization in which each part understands only its only narrowly defined task. Hedlund acknowledged that such holographic qualities were 'only a theoretical ideal' (1986: 24) but his argument nonetheless reinforces a critical point: that large MNCs need to become much better at managing the internal flow of knowledge between units, in order to leverage their geographically dispersed capabilities.

In this chapter we report on an empirical investigation into the geographical dispersal of market knowledge in large MNCs. The study addresses two questions, one applied and one theoretical. The applied question, which any senior manager in an MNC can relate to is: Where are our best practices or leading-edge marketing capabilities? This question recognizes that there are pockets of knowledge distributed throughout the corporation, but that identifying and making use of such capabilities is a far from trivial task. While a number of recent studies have looked the internal transfer of knowledge in MNCs (Kostova and Cummings, 1997; Szulanski, 1995; Zander and Kogut, 1995), our focus is on the antecedent question, namely, where is the knowledge that the MNC wants to transfer?

The theoretical question is 'Why do we see a significant variation in marketing capabilities across different subsidiary units?'[1] Our argument here is essentially that the MNC can be modelled as a differentiated network (Ghoshal, 1986; Nohria and Ghoshal, 1997) in which each subsidiary unit takes on a unique form, and develops unique capabilities, to reflect the demands of its specific task environment. Thus, the variation in marketing capabilities across subsidiaries is a function of the variation in external environments they face. At the same time, transfer of knowledge within the firm is 'sticky' (Szulanski, 1995; von Hippel, 1994), with the result that transfers of marketing capabilities across the firm (Quelch and Hoff, 1986) are relatively rare. In other words, capabilities do not flow easily between subsidiary units. It is therefore the combination of external differentiation and sticky internal knowledge flows that leads to the emergence and perseverance of the variation in marketing capabilities.

In the body of the chapter we play out these arguments in greater detail, and use them to set out a number of propositions relating various aspects of the subsidiary unit's external environment and its relationships with the rest of the MNC to its observed marketing capabilities. In the empirical part of the chapter we describe and report on a study of 115 subsidiary units in six Swedish MNCs. The data are used to test the propositions. We then discuss the implications of our findings for the theory of the MNC and for practice.

Background and conceptual framework

There are two distinct schools of thought on the theory of the multinational corporation.[2] The first line of thinking, emerging from trade economics, saw the *raison d'être* of MNCs as their ability to internalize transactions across national boundaries (Hymer, 1976). Many variants of this approach have been put forward (e.g. Buckley and Casson, 1976; Dunning, 1980; Rugman, 1981) but they are all concerned with explaining *why* MNCs exist in the first place, and they all build on transaction cost economics. This line of thinking leads us to emphasize the necessity and benefits for MNCs in internalizing knowledge transactions across national boundaries.

The other line of thinking is based in organization theory, and is concerned much more with *how* the MNC works rather than why it exists. The distinctive feature of the MNC, according to this approach, is that it operates in multiple countries, each of which is characterized by a distinct task environment or organizational field (Ghoshal and Nohria, 1989; Westney, 1993). In order to respond effectively to its environmental heterogeneity, the MNC must differentiate the activities of its subsidiaries (i.e. it must allow them to adapt to their immediate task environment) but it must also integrate them (i.e. build internal relationships to enable cooperation and learning). While the differentiation–integration dichotomy was first put forward by Lawrence and Lorsch (1967) to explain a simple functional organization, it has since become very widely used in the MNC literature (Prahalad and Doz, 1987; Bartlett and Ghoshal, 1989) because of the extreme level of environmental heterogeneity the MNC faces.

Our focus in this chapter is on the latter theory, which we will refer to as the differentiated network model of the MNC (Nohria and Ghoshal, 1997). As with the transaction-cost approach, there are many variants here including Ghoshal and Bartlett's (1990) interorganizational network, Hedlund's heterarchy (1986, 1993) and N-Form organization (1994), and Westney's (1993) institutional theory of the MNC. However, rather than getting into a review of the specific arguments and relative merits of these contributions, our approach here is to focus our discussion around the two key elements of the differentiated network model, namely (a) adaptation to the external environment and (b) integration of activities inside the firm.

Adaptation to the external environment

The open-systems perspective on organization theory sees the focal organization adapting to and interacting with other organizations in its task environment (Katz and Kahn, 1966; Thompson, 1967). In the case of the MNC that is operating in multiple countries, it thus follows that each subsidiary unit has to adapt to its own unique task environment consisting of local suppliers, customers, competitors and governmental bodies (Ghoshal and Bartlett, 1990). The activities that the subsidiary undertakes, and the capabilities that it develops over time, will be a function of the specific demands placed on it by actors in the local environment.

There has been a considerable amount of research in recent years that has built explicitly or implicitly on the differentiated network model. Careful examination of this research, however, reveals that the question of *what* is being differentiated varies from study to study. Three groups of studies can be discerned. The first group looked at the *actual activities* performed by the subsidiary. Bartlett and Ghoshal (1986) for example, identified four different roles, such as strategic leader, black hole and implementer that varied according to the strategic importance of the local market and the strength of the subsidiary. Jarillo and Martinez (1990), in similar fashion, identified strategic roles as a function of the demands of the subsidiary's industry. The second group looked at the *structural context* of the subsidiary, i.e. the nature of its relationship with HQ and with other units. Ghoshal and Nohria (1989) for example, found that the use of various HQ-subsidiary control mechanisms varied according to the uncertainty and resource scarcity in the local environment (cf. Lawrence and Dyer, 1983), and Nobel and Birkinshaw (1998) showed a link between the strength of the host country network and the relationship of the R&D subsidiary to its corporate network. Finally, the third group looked at *management practices* in subsidiary units, and related them to the competing demands of the host country and the corporate parent (e.g. Rosenzweig and Nohria, 1995), typically finding that the host country exerted a stronger pull than the parent firm.

The distinction between the first and the third of these groups is very important in the context of the overall theory of the MNC. Differentiation of activities is a form of task specialization, the result of which could be that all R&D in semiconductors is performed in California while all manufacturing of semiconductors is performed in South Korea. These activities are then leveraged globally, without any other subsidiary units around the world being involved. This task specialization is generally hierarchically implemented.

Differentiation of management practices, by contrast, relates to the way that certain common tasks are performed. Thus, even though all marketing subsidiaries around the world have to undertake after-sales service for example, it is possible that the demands of the American marketplace will force the US subsidiary to develop a particularly strong set of capabilities in after-sales service. Unlike the actual activities performed by the subsidiary, which typically involve specialized assets and some form of corporate 'charter', management practices are usually intangibles that can be disseminated and/or transferred to other subsidiaries at little or no cost to the source subsidiary. In other words, the management practices are generally developed in a particular subsidiary in a particular local environment, and could then be used in other subsidiaries in other local environments.

To anticipate a later section, this study is concerned with the differentiation of practices in the marketing area, and thus with the dissemination (or lack thereof) of those practices to other parts of the corporation. The benefits of multinationality, in other words, arise from the sharing and transfer of good marketing ideas across locations (Quelch and Hoff, 1986), rather than global standardization of the elements of the marketing mix. Moreover, it is important to be clear that marketing capabilities have both a generic and a market-specific component. The French subsidiary could, for instance, have a very strong PR department through its excellent ties to the local media. However, this can be the result of a unique approach to managing media relationships, or it can be because the PR head went to school with the heads of the radio and TV stations. Clearly, the former capability is generic, and potentially transferable to other subsidiaries, while the other is location-specific, and of no value outside France.

Integration of activities inside the firm

The second part of the differentiated network model is the way that the various subsidiary units are brought together, so that the whole is worth more than the sum of the parts. This is typically referred to as *integration,* which Lawrence and Lorsch (1967) defined as the quality of the state of collaboration among departments (read subsidiaries) as well as the techniques used to achieve this collaboration.

It is possible to identify two rather different lines of thinking on the integration of activities in MNCs. The first comes from the work of Prahalad (1976), Prahalad and Doz (1981), Bartlett and Ghoshal (1989) and Ghoshal and Bartlett (1994), and it may be referred to as the organization context approach. Here the logic is that integration occurs through the creation of an organizational context that facilitates proactive cooperation between subsidiaries. Rather than resorting to simple structural solutions, the organization context approach argues that top management has to create evaluation and reward systems, information systems and a social structure that fosters the required behaviours among employees.

The second line of thinking is concerned explicitly with the transfer of knowledge inside large MNCs. Obviously this is a narrower perspective that is to some degree subsumed by the first approach, but it also provides some important insights especially when it comes to the transfers of knowledge. Strictly speaking this line of thinking can be traced back to the technology transfer literature (e.g. Davidson, 1980; Mansfield and Romeo, 1980; Teece, 1976), but of much greater interest here are the recent studies of internal knowledge transfers by Kostova and Cummings (1997), Szulanski (1995) and Zander (1991). Zander (1991) made the important contribution that the characteristics of the knowledge being transferred (e.g. how articulable it is) have a big impact on the effectiveness of internal transfer and external imitation. Szulanski (1995) used communication theory (Shannon and Weaver, 1949) to show that the 'stickiness' of knowledge can be modelled in terms of the attributes of the source, the recipient, their relationship, and the context in which they interact. Kostova and Cummings (1997) showed the importance of institutional factors (at a country and organizational level) on the effectiveness of cross-border transfers of management practices.

Overall, there are great many ways that the integration of activities within the firm can be achieved, but of course what is important here is that *integration is far from perfect*. If in fact these integration processes were efficient, we would expect to see leading-edge management practices flow rapidly and seamlessly from country to country. Yet there is strong evidence in the literature (Chew *et al.*, 1990; Leibenstein, 1966; Szulanski, 1995), as well as through casual observation, that this does not happen.

Why is the transfer of management practices not more efficient? A partial answer that Szulanski (1995) argued very persuasively is that knowledge assets are inherently 'sticky' due to the nature of the knowledge in question (its causal ambiguity), and the attitudes of the source and recipient (their lack of motivation, etc.). This approach suggests that even when the relevant practice, the source and the recipient have all been identified, the effective transfer is far from assured. The 'not invented here' (NIH) syndrome (Katz and Allen, 1982) is a well-known example of the sorts of behavioural pathologies that get in the way of apparently rational knowledge transfer processes.

Our approach in this research is to draw attention to a slightly different reason for the inefficient transfer of management practices, *namely, the lack of knowledge about the whereabouts of leading-edge practices*. Here, we are starting with the premise that managers are boundedly rational, and gravitate towards satisficing solutions to their problems (Cyert and March, 1963). Given the size and heterogeneity of large MNCs, and the inherent stickiness of knowledge, we argue that the fundamental problem the MNC faces is the identification of the appropriate practice to transfer. The 'holographic characteristics' (Thompson and Tuden, 1959) that Hedlund (1986, 1993) referred to, in other words, are still a distant ideal. Of course, the question of how to achieve an effective transfer is no less important, but it would appear to be secondary to the identification of the appropriate practice. The focus on the problem of identifying units' abilities builds on a framework where insight into units' actual ability is a pre-requisite for effective and systematic use of a firm's dispersed capabilities. To our minds this approach has not been thoroughly addressed in IB research. Figure 8.1 illustrates our approach.

Figure 8.1 indicates the two necessary steps that are needed effectively to utilize the MNC's globally dispersed resources. The first step is

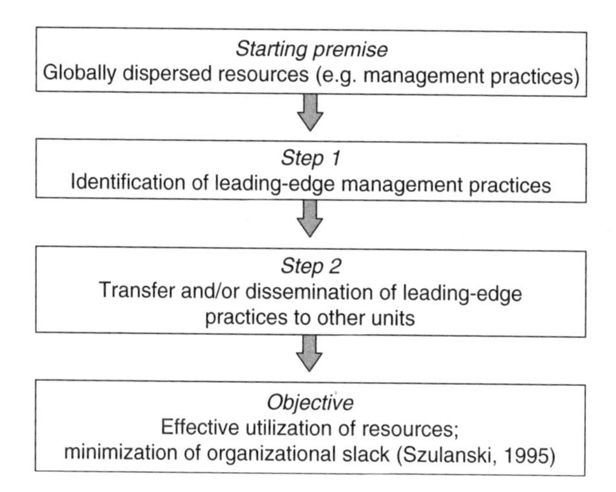

Figure 8.1 Steps in utilization of an MNC's resources

simply identifying where the leading-edge practices are; the second step is evaluating and implementing the transfer opportunities that are identified. The outcome of the two steps is an enhanced utilization of the MNC's globally dispersed resources, or in Szulanski's (1995) terms, a reduction in organizational slack (i.e. because resources are being more efficiently used). From the discussion above it should be obvious that this paper is focused exclusively on Step 1. Other work discussed above, and a future piece of work from the current project, examine Step 2 in greater detail.

One more comment on Figure 8.1 is necessary, because we want to make it clear that the sequential framework is designed for analytical clarity and *not* to represent the inner workings of the MNC. In fact, in our experience the transfer process is often undertaken before any efforts have been made to find out whether the chosen source is actually a best practice. The reality, then, is something akin to a garbage-can decision model (Cohen *et al.*, 1972) in which currently available and known resources are combined, transferred or leveraged to solve immediate and local problems, but without any great concern for their global, system-wide utilization.

Marketing capabilities

As stated earlier, the focus in this study is specifically on the marketing capabilities of the sales and marketing subsidiaries of large MNCs.

The guiding research question can thus be stated as follows: *What factors (external and internal to the firm) influence the level of marketing capabilities in an MNC subsidiary?* Our decision to study marketing capabilities was based on the simple observation that they are the most geographically dispersed capabilities in the firm.[3] Moreover, this is to our knowledge the first study that has explicitly focused on marketing activities in MNC subsidiaries, in contrast to manufacturing and R&D activities which have been studied many times over (see Birkinshaw and Morrison, 1995; Birkinshaw *et al.*, 1998; Håkanson and Nobel, 1993; Kümmerle, 1996).

The marketing literature, interestingly, has not much to say about the variation in marketing capabilities across MNC subsidiaries. Marketing activities with low degrees of local market specificity, i.e. not being usable only in one local market and potentially being organized in a similar fashion across markets and intra-organizational boundaries, have been described as international, multinational, multi-regional and/or global marketing activities (Levitt, 1983; Albaum *et al.*, 1989; Kotler, 1994; Cateora, 1993; Keegan, 1989). These different typologies of marketing strategies may be subsumed under the umbrella category of international marketing (Jeannet and Hennessey, 1995). The common denominator in this work is the focus on marketing functions which are executed in more than one country simultaneously. However, this literature tends to downplay organizational issues, and it also tends to work on the implicit assumption that the firm is a monolithic entity. As a result, the question of internal transfers of marketing practices between subsidiaries has not been considered.

Where the marketing literature did prove useful, however, was in its existing measures of various marketing capabilities. In particular, we decided to focus on a construct called *market orientation*, which is defined as the organization-wide generation of market intelligence, dissemination of the intelligence across departments and organization-wide responsiveness to it (Kohli and Jaworski, 1990). While this construct was originally used at another level (Jaworski and Kohli, 1993), we decided to use it at a subsidiary level, using the logic stated above that market orientation is likely to vary significantly from subsidiary to subsidiary. In addition, we measured a number of other marketing capabilities at the subsidiary level that were also taken from the marketing literature.

Proposition development

We are now in a position to specify the propositions that will be tested in this research. Figure 8.2 indicates the overall framework, with a series

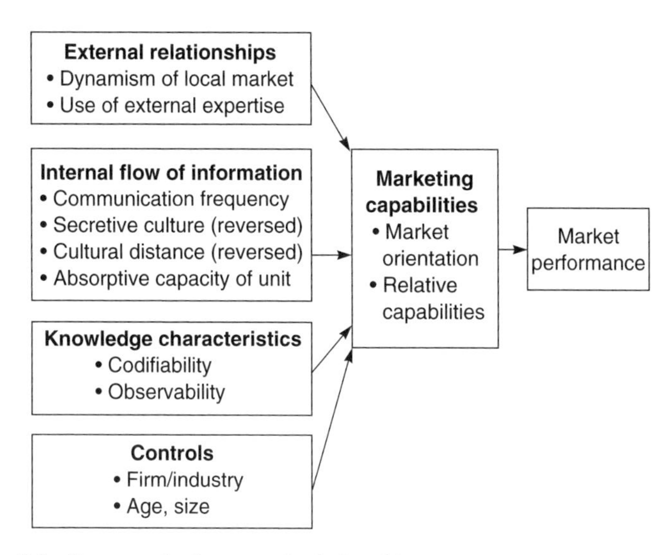

Figure 8.2 Framework of proposed relationships

of constructs predicting the level of marketing capabilities in the subsidiary, and one proposition linking subsidiary marketing capabilities to subsidiary market performance.

Differentiation of marketing capabilities

As we have argued, every individual marketing subsidiary will adapt over time to the specific conditions and requirements of its local market. The network perspective, in particular, argues that capabilities emerge through interaction with key suppliers and customers in a process of mutual adaptation (Forsgren and Johanson, 1992). Considering the MNC as a whole, the extent to which each marketing subsidiary operates in a distinct task environment with its own unique set of relationships, will therefore be strongly related to the emergence of a differentiated network of marketing capabilities.

At the level of the individual subsidiary, we argue that certain sets and types of relationships will *ceteris paribus* be associated with higher marketing capabilities. The first set relate to the perceived dynamism of the local marketplace, as defined in Porter's (1990) *Competitive Advantage of Nations* study (cf. Sölvell *et al.*, 1991). Using this approach, it is argued that the more local competition, the greater the strength of related and supporting industries and the strong the demand factors, the more competitive (broadly defined) the firm becomes. Translated in to marketing terminology, we suggest that dynamism is likely to lead to greater marketing capabilities in the subsidiary.

> **Proposition 1**: The perceived dynamism of the local market will be positively correlated with the marketing capabilities of the subsidiary.

The second set of network relationships we look at are those that provide a source of expertise for the subsidiary. Thus, if proposition 1 is cast in general terms about the importance of market dynamism, the second looks specifically at those relationships that the marketing subsidiary draws on for expertise. The proposition is simply that the more the subsidiary uses external expertise (from various sources), the greater its marketing capabilities will be.

> **Proposition 2**: The extent to which external expertise is used by the subsidiary will be positively correlated with the marketing capabilities of the subsidiary.

We should note here that we plan to identify a number of additional local market constructs that might be expected to predict different levels of marketing capabilities, including the size of the local market and the strategic importance of the local market. The necessary data for these constructs have not yet been collected.

Barriers to and facilitators of internal transfer of practices

The second sets of factors are those that *impede* the internal dissemination of knowledge within the MNC. Our argument here is slightly different from that presented earlier, because we are now trying to explain why certain subsidiary units will have strong marketing capabilities. Essentially, we propose that efficiency of internal knowledge flows in the MNC will vary from place to place: some subsidiaries will be well-connected to each other, and may exhibit a relatively efficient flow of practices, while others will be relatively isolated and thus exhibit much greater stickiness in knowledge flows.

Without claiming to have identified all the key dimensions of internal knowledge flows, we suggest that four sets of factors are particularly relevant. First, the absorptive capacity of the focal subsidiary is hypothesized to be related to the level of marketing capabilities, because it is indicative of the subsidiary's ability to make sense of, interpret and apply new ideas that it picks up from elsewhere (Cohen and Levinthal, 1990).

Absorptive capacity is typically used to proxy the ability of the receiving unit to facilitate a knowledge transfer (Szulanski, 1995), but here we used it more broadly to refer to the subsidiary's ability to identify and learn from marketing capabilities throughout the MNC.

> **Proposition 3**: The absorptive capacity of the subsidiary will be positively correlated with the marketing capabilities of the subsidiary.

Second, we argue the level of communication between the subsidiary and other parts of the MNC will be indicative of the extent to which knowledge is being shared (Hedlund, 1993; White and Poynter, 1990), using the simple logic that transferring capabilities is a highly communication-intensive process. Obviously this does not get at *what* is being communicated, so there is likely to be a great deal of noise in the measure, but on the other hand it is a behavioural measure (i.e. how often do you communicate with HQ) rather than an attitudinal one, so it might be considered more valid on those grounds.

> **Proposition 4**: The frequency of communication between the subsidiary and other internal actors will be positively correlated with the marketing capabilities of the subsidiary.

We also identify two sets of inhibitors of knowledge flow. First, we argue that some corporations have a more 'secretive' culture than others. This is manifested in a reluctance to share ideas, to hoard information, and not to attempt to learn from one's sister units. Of course, one could argue that this is likely to be a corporate-wide phenomenon, but our approach here is to argue that on such questions perception is reality, and that perceptions are likely to vary from subsidiary to subsidiary. Given our earlier arguments about the stickiness of knowledge, it also seems quite likely that different parts of a single MNC could have different approaches to secrecy. Thus, we suggest simply that the extent to which subsidiary management perceives the culture to be secretive will be negatively correlated with the marketing capabilities of the subsidiary.

> **Proposition 5**: The extent to which subsidiary management perceives there is a secretive culture in the firm will be negatively correlated with the marketing capabilities of the subsidiary.

Finally, we argue that cultural distance from the head office, measured in terms of Hofstede's (1991) four dimensions, is likely to be negatively related to the marketing capabilities of the subsidiary. The logic here is that cultural differences represent a barrier to the flow of knowledge because the message is less easily communicated. Given that a large percentage of the valuable knowledge will come from or through head office, it follows that the more distant (culturally speaking) the subsidiary is from head office, the weaker its marketing capabilities will be. Thus:

> **Proposition 6**: The cultural distance between the subsidiary's country and the MNC's home country (Sweden) will be negatively correlated with the marketing capabilities of the subsidiary.

Characteristics of knowledge

Related to the question of internal transfer are the characteristics of the knowledge itself. Here we use the well-established distinction between tacit and codified knowledge (Polanyi, 1969) to argue that the more tacit knowledge is the more difficult it is to transfer it effectively from place to place (whether inside or outside the firm). This argument is supported by the work of Zander (1991) and Szulanski (1995) (though he used the term causal ambiguity, not tacitness). Moreover, following Zander (1991) and Winter (1987) we also identified the concept of observability, which refers to the extent to which others can make sense of the knowledge by observing it in use. Both codifiability and observability are *ceteris paribus* argued to lead to more easy transfer and thus stronger marketing capabilities in the subsidiary.

> **Proposition 7**: The extent to which the subsidiary's knowledge base is easy to codify and observe will be positively correlated with the marketing capabilities of the subsidiary.

Firm level effects

Finally, it is important to acknowledge the potential importance of a firm-level effect in our analysis. Because the data covers 115 subsidiaries in six MNCs, it is possible that the single biggest determinant of differences in market capabilities would be the MNC to which the subsidiaries belong. If so, this would provide considerable support for the marketing literature which tends to treat firm marketing orientation as a homogeneous construct (Jaworski and Kohli, 1993), and indeed find against the network model of the MNC which sees subsidiary capabilities as differentiated. Our position here is midway between these positions, in that we expect to see some variance explained at the firm level, but probably less than by the other predictor variables. In order to be more specific here, we argue that those subsidiaries that are in industrial products MNCs are likely to have somewhat weaker marketing capabilities than those in consumer products MNCs because industrial products are sold more on their technical qualities than according to the skill of their marketers.

> **Proposition 8**: There will be a significant firm-level effect, namely that subsidiaries in industrial products MNCs will have weaker marketing capabilities than those in consumer products MNCs.

Market performance

The final proposition we put forward is that we would expect marketing capabilities to be positively related to performance, which is here defined as the market performance (sales, market share, profitability) of the subsidiary (Jaworski and Kohli, 1993). It is also important to note that we do *not* anticipate a relationship between any of the independent variables and market performance directly, only through marketing capabilities.

> **Proposition 9**: The marketing capabilities of the subsidiary will be positively correlated with the subsidiary's market performance.

Research methodology

The empirical findings presented in this chapter are part of a multi-phase study called 'Best Practices in Market Organizations'. This research is being conducted in close cooperation with six Swedish multinational corporations. At the time of writing about 80 per cent of the total data had been collected. The rest will be collected in the first few months of 1998. What follows is a brief overview of the data collection process.

The first phase of research was a detailed set of interviews in one MNC with a smaller number of interviews in four others. During this research we picked up on the emergence of 'best practices' in certain marketing subsidiaries, such as the 'key account management' practice in a Swedish subsidiary of an American MNC. This then led us to focus our attention on the wide variation in marketing capabilities from one country to the next, even within the same MNC, and thus to an analysis of the factors that could lead to such variations.

In the second phase of the research we put together a draft questionnaire which was presented, discussed and then pilot-tested in one Swedish MNC. The questionnaire was based on a mixture of established scales from the literature and our own measures of constructs that appeared to be very relevant in this context. This questionnaire was sent to 19 managers in five countries, 17 of whom completed and returned it. In addition we got three managers at the corporate headquarters to fill in a slightly different questionnaire, in which they assessed the marketing capabilities of the five country operations. The results from the pilot survey were instructive. Some of the scales we had developed ourselves had to be changed, which led us to drop several items when the pilot-test answers proved ambiguous. In addition we were able to test the inter-rater reliability of the questionnaire by comparing the answers from the multiple respondents at the same units. Overall the inter-rater reliability was acceptable given that the within-group variance was lower than the between-group variance. We were surprised, however, by the extent to which HQ managers' assessments of units' capabilities and units' own assessments of their capabilities differed. There is not space here to elaborate on this point, and at the current time we have not received the HQ responses for any other firms, but our preliminary expectation here is that perception gaps between HQ and subsidiary units are *very significant*, and that maybe their size can itself tell us something interesting about the inner workings of the MNC. In our future steps in this project, we will devote much attention to the issue of 'how much does the firm know about what it knows'.

In the third phase of research we broadened our investigation to a further five Swedish MNCs. In each one, we first approached corporate management to get permission to do the research, and having enlisted their support we got the lead contact (the global marketing manager or equivalent) to provide us with a list of all sales and marketing subsidiaries around the world, and the president/managing director of each (to whom the questionnaire would be sent). The five companies were selected on a convenience basis, but with a view to sampling across a wide variety of industries. We also approached a sixth company, which declined to participate, and a seventh company, which decided to participate. The subsidiary data from the seventh company will be collected by the end of 1997. Table 8.1 lists the participating companies, their principal industry, and the number of sales and marketing subsidiaries that were polled. Overall the response rate for the survey was 73 per cent, though this will rise slightly in the next couple of months as the last few questionnaires arrive.

Table 8.1 Information on sample

Firm/industry	*Approximate turnover 1996 (MUSD)*	*No. of sales and marketing subsidiaries*	*No. of responses received*	*Response rate (%)*
Insurance	3200	5	5	100
Steel	1300	27	27	100
Tools	1300	31	29	94
Electronics	3250	46	27	59
Automobiles	11850	29	21	72
Pharmaceuticals	7200	26	18	69
Total		157	115	73

It should also be noted at this stage that the questionnaire was tailored to some degree to each company in question. Many questions were altered slightly to reflect the terminology of the company, e.g. product divisions *vs* business areas *vs* market areas. In addition we also created a number of firm-specific questions that targeted specific marketing practices of each company. However, none of these were used in the current analysis.

Construct operationalization

The majority of questions were attitudinal, in that they asked respondents to assess the extent to which they agreed with the question on a 1–7 Likert scale. In addition we also asked a number of factual questions, such as the subsidiary's local market share or its number of employees. Finally, we also collected some data from secondary sources, such as the size of the local market, and the geographical and cultural distance from headquarters.

Marketing capabilities. As noted earlier, we operationalized marketing capabilities in two very different ways. First, we measured the subsidiary's *market orientation* [MO][4] using Jaworski and Kohli's (1993) established index. Because of space constraints we ended up shortening their instrument (by deleting those items that had the weakest loading in the pilot questionnaire), leaving 21 items. Altogether this scale had very high reliability (Cronbach's Alpha = 0.83). Of course, it would be possible to split this scale into its constituent parts (collecting, distributing, responding to market information) but at this preliminary stage we decided to use it as a single construct.

Second, we asked respondents to assess their *relative marketing capabilities* [Relmkcap], i.e. relative to other subsidiaries in the same firm. This question had five dimensions: collecting market information, distributing market information, analysing and acting on market information, linking up with customers' value-chain and general marketing, where 1 was 'much worse' and 7 was 'much better'. This scale also had very high reliability (Alpha = 0.87), though interestingly it had a fairly modest correlation to market orientation (r = .31). Finally, we also specified a number of firm-specific questions about marketing capabilities that the HQ managers requested be included. This four-item scale (with different questions in each company) had a very strong correlation with relative marketing capabilities (r = .73), so we simply elected to drop it in the current analysis.

Market performance [Relmkper]. This was a four-item scale in which we simply asked respondents to rate on a 1–7 scale their relative performance along four dimensions: overall sales revenue, sales revenue growth, overall market share and operating profit. While these are all very different measures of performance they correlated quite strongly with one another so we summed them to form a single measure. The reliability was moderate (Alpha = 0.65).

Use of external expertise [Extexp]. We developed a scale specifically for this study to understand the sources of expertise on which the subsidiary draws. Respondents were asked the extent to which they agreed with the following questions: (a) We frequently draw on external expertise when we perform our activities; (b) We regularly draw on critical expertise from other <firm> companies; (c) Country/BA functions are providing vital expertise for our daily activities; and (d) Other <firm> local companies are important sources of expertise for us. While these questions were obviously getting at different sources, we were able to create a single scale with moderate reliability (Alpha = 0.55).

Dynamism of local market [Mktdyn]. Based on questions developed by Woodcock (1994) and used in Birkinshaw *et al.* (1998), we asked respondents to answer questions on four aspects of their local market: (a) Competition in the local market is extremely intense; (b) Relationships between our company and suppliers/buyers are very strong; (c) The speed of product or service innovation by competitors is high; and (d) Market demand is growing rapidly in our local market. Again, these four questions addressed rather distinct facets of the local market but we were able to construct a scale with moderate reliability (Alpha = 0.50). To capture fully aspects of dynamism and use of external expertise in the local market, we will include factors such as market size and types of customers in our next step of this analysis.

Communication frequency [Commun]. Based on a simple frequency scale where 1 = daily and 7 = yearly or less (see Ghoshal, 1986; Nobel and

Birkinshaw, 1998), this scale asked respondents to indicate often how they communicated with (a) HQ managers, face-to-face and through other means, and (b) managers in other local companies, face-to-face and through other means, to discuss operations. The four questions together gave a single scale of adequate reliability (Alpha = 0.74). In this analysis, we have kept the answers as they are but we will reverse-code the communication items in a later stage so that a higher number is associated with more frequent communication rather than vice versa.

Secretive culture [Secret]. A three-item scale was developed to measure the extent to which the firm had a secretive culture. Questions as follows: (a) People are very secretive about their ideas in <firm>; (b) I would get no recognition at all in <firm> if I helped another local company; and (c) The 'not-invented-here' syndrome is a real problem in <firm>. Reliability for these items was moderate (Alpha = 0.68). The scale was reverse-coded (i.e. a 1 was reversed into a 7, a 2 into a 6 and a 3 into a 5), which actually means that we measure lack of secretive culture.

Cultural distance [cultdist]. We used Kogut and Singh's (1988) measure which estimates the euclidean distance from the subsidiary country to the HQ country (Sweden) using Hofstede's (1991) dimensions of national culture.

Absorptive capacity of subsidiary [Absorb]. Based on the Cohen and Levinthal's (1990) concept of absorptive capacity and in part on Szulanski's (1995) measures, we put together our own 5-item scale as follows: (a) <Firm> has a shared language that helps us learn from other local companies; (b) We have a clear vision of how <firm> companies could help us; (c) We know which activities we perform poorly and which to improve; (d) We have the ability to effectively absorb capabilities from other companies; (e) We almost always know where and how to search for ways to improve the way we perform certain activities. Reliability for these items was adequate (Alpha = 0.71).

Codifiability of knowledge [Codified]. We took Zander's (1991) scale which was also used in Nobel and Birkinshaw (1998). Respondents answered the following questions about their knowledge base: (a) A manual describing how our activities are executed could be written; (b) New staff can easily learn how to perform the services that our local company offers by talking to skilled employees; (c) Training new personnel is typically a quick and easy job for us; (d) New personnel with a university education can perform the services that our local company offers. Reliability for these items was moderate (Alpha = 0.67).

Observability of knowledge [Observe]. Again, Zander's (1991) scale was used, though only two items hung together well: (a) It's easy for local companies to understand how we perform our activities; (b) Large parts of our products and services are embodied in methodologies that are easily adapted by other <firm> local companies.

Controls. We used the following control variables: *Subsidiary age* (the year of founding [Age]), *subsidiary size* (number of employees in total [Employees]) and *dummy variables* for the companies in the study [Steel, Tools, Electronics, Autos and Pharmaceuticals (or Pharma)]. In a later stage, we will extend the list of control variables and also develop the items in coherence with previous studies such as Narver and Slater (1990), Jaworski and Kohli (1993), and Slater and Narver (1995).

Findings and discussion

The propositions were tested using multiple regression analysis. Table 8.3 presents the findings from the analysis. Models 1 and 2 examine the predictors of marketing capabilities for two different constructs (market orientation, relative marketing capabilities), and using two different methods in each case (all variables entered, stepwise). Model 3 examines the predictors of market performance. Table 8.2 lists the correlations between all constructs.

The correlation table (Table 8.2) provides provisional support for most of the propositions. Market orientation is significantly correlated with all of the independent variables except market dynamism. Relative marketing capabilities is significantly correlated with market dynamism, communication and knowledge codifiability. In addition we find that one of the firm dummies is significant in each case. Table 8.2 also shows that there are significant correlations between many of the independent variables, especially those related to internal knowledge flows. However, these correlations are low enough that multicollinearity was not deemed to be a major problem.

Table 8.3 indicates the results of the regression analysis. Considering Model 1 first, market orientation is predicted by five constructs: (1) absorptive capacity of the subsidiary (proposition 3); (2) frequency of communication (proposition 4); (3) a percieved lack of secretive culture (proposition 5); (4) the perceived codifiability of knowledge (proposition 7), and (5) the electronics firm negatively, i.e. subsidiaries from that firm had lower market orientation (proposition 8). Overall, 39 per cent of the variance in market orientation was explained by these five constructs, a very satisfactory result. For Model 2

Table 8.2 Correlation matrix

	MO	Relmkcap	Extexp	Mktdyn	Commun	Secret	Absorb	Cultdist	Codified	Observ	Relmkper	Steel	Tools	Electronics	Autos	Pharma	Age	Employee
MO		.313**	.251**	.172	−.232*	.458**	.522**	−.199*	.227*	.294*	.153	.063	.047	−.223*	.130	−.034	−.173	−.081
Relmkcap			.111	.206*	−.197*	.050	.115	−.187	.215*	−.027	.375**	−.021	−.321**	.158	.117	.143	.052	−.010
Extexp				.200*	−.257**	.206*	.378**	.011	−.150	.279**	−.093	−.166	−.003	.092	.119	−.850	.035	.042
Mktdyn					−.096	.113	.294**	−.104	.049	.072	.023	−.247**	−.019	.222*	.002	.020	.189	.100
Commun						−.002	−.223*	.044	.016	−.130	−.074	.203*	.073	−.228*	−.126	.088	.186	.022
Secret							.392**	−.158	.002	.360**	.014	.188*	.062	−.078	−.095	−.107	−.206*	.033
Absorb								−.116	.196*	.476**	.009	.090	.074	−.076	.014	−.194*	−.098	.016
Cultdist									−.185	−.014	−.125	.095	.046	.024	−.021	−.128	.154	−.068
Codified										.263**	.119	−.030	−.015	−.061	.072	.041	.016	−.115
Observ											.084	.078	.103	−.093	.094	−.251**	−.267**	.004
Relmkper												−.114	−.067	.182	.011	.089	−.050	.051
Steel													−.318**	−.289**	−.228*	−.196*	−.172	−.142
Tools														−.303**	−.239**	−.205*	−.222*	−.054
Electronics															−.217*	−.186*	.230*	.277**
Autos																−.147	.019	−.083
Pharma																	.149	.043
Age																		−.306*
Employee																		

*** $p = .001$; ** $p = 0.01$; * $p = 0.05$.

Table 8.3 Regression analysis of Models 1, 2 and 3

Independent variables	*MODEL 1 Dependent variable: Market orientation*		*MODEL 2 Dependent variable: Relative marketing capabilities*		*MODEL 3 Dependent variable: Performance*	
	Stepwise	*Enter*	*Stepwise*	*Enter*	*Stepwise*	*Enter*
	Standardized Beta coefficients		*Standardized Beta coefficients*		*Standardized Beta coefficients*	
Constant						
Absorb	.296***	.284**		.009		−.002
Secret	.325***	.336***		.040		−.051
Commun	−.213*	−.170†	−.259**	−.231*		.096
Codified	.159*	.174†	.227*	.264*		−.058
Extexp		.115		.154		−.153
Mktdyn		.069	.168†	.147		−.127
Observ		−.105		−.075		.139
Steel		.000		.092		.020
Tools		−.045	−.228*	−.117		.192
Electronics	−.232**	−.191		.079		.494*
Automobiles		.106		.132		.208
Pharma		.081		.192		.200
Age of unit		−.095		−.063		−.204
Employees		−.089		.045		−.166
Market orientation						.126
Relative marketing capabilities					.375***	.352**
Adjusted R^2	.391	.373	.181	.141	.132	.138
F	13.701	5.207	6.398	2.152	15.557	1.958
Significance	.000	.000	.000	.016	.000	.026

*** $p = .001$; ** $p = 0.01$; * $p = 0.05$; † $p = 0.1$

the dependent variable was the relative marketing capabilities of the subsidiary. This model explained a lower percentage of the variance in marketing capabilities ($R^2 = 0.18$), using four predictor variables in the stepwise model: (1) the dynamism of the local market (proposition 1); (2) the frequency of communication (proposition 4); (3) the codifiability of the marketing knowledge (proposition 7); and (4) the tools firm negatively, i.e. subsidiaries from that firm had lower marketing capabilities (proposition 8).

Taken together, Models 1 and 2 suggest a number of important conclusions. First, the evidence suggests that the internal flow of knowledge (or its absence) is generally a more important determinant of the subsidiary's marketing capabilities than the heterogeneity of the external environment. In particular, the perceived codifiability of knowledge is a significant predictor variable in both models, as is the level of communication between the subsidiary and other parts of the MNC. Of course, this finding is probably related to the fact that we have not yet collected a comprehensive set of local market variables (see earlier), but it is at least interesting to see that the internal knowledge flow characteristics are important determinants of marketing capabilities. A second important finding is that we do observe a firm effect but not a very strong one. In both models there was one firm dummy significant. Moreover, by inspecting the correlation matrix in Table 8.2 we see that our expectation regarding consumer *vs* industrial products is broadly confirmed – the automobile and pharmaceuticals firms come out with higher capabilities on average, and the tools, steel and electronics firms come out lower. The one exception is the electronics firm, and this is a case of a company with a strong engineering background that has only recently moved into consumer electronics.

Model 3 provides strong support for proposition 9, namely that market performance will be predicted by marketing capabilities and nothing else. This finding, however, is only true for the relative marketing capabilities construct, and not the market orientation of the subsidiaries. This is contrary to Jaworski and Kohli's (1993) finding. It is also somewhat difficult to interpret because the two marketing capability constructs are strongly related to one another. Finally, we should also note that there was one other predictor of market performance, namely the electronics firm dummy. This result is interesting since this firm also proved to have much lower market orientation than the other firms (see above). This result suggests that market orientation is too focused on internal organization of externally focused activities, i.e. customer orientation, and does not adequately cover local-specific development of customer relations or internal, intra-departmental relations.

Limitations and conclusions

In terms of the limitations of the current chapter, we should first acknowledge that it is work-in-progress. We will add a number of additional variables about the local marketplace, and a headquarters assessment of the subsidiaries' relative marketing capabilities. We also expect about 30 or 40 additional questionnaire returns. Once all of this has been done, we will have a much more definitive set of findings regarding the determinants of subsidiary marketing capabilities.

This chapter also hints at a number of areas for future research. First, the market orientation construct has been criticized for being too focused on customer orientation (Sharp, 1991). We will consequently build constructs that more precisely pick up on the linkages between the firm's technological activities, the market orientation of the sales and marketing activities in addition to building relations with customers' value-chains. Second, we will explore and elaborate on the question of 'holographic' characteristics (Thompson and Tuden, 1959; Hedlund, 1986, 1993). From what we have observed so far, the real world is a long way from the 'ideal' of a holographic corporation, so we intend to look more closely at the extent to which the MNC 'knows what it knows' by comparing our subsidiary and headquarters level data. It may, in fact, be that the costs of holographic qualities outweigh the advantages, and this is something that our data should be able to address.

To conclude, this chapter makes what we believe to be two important contributions. First, we provide some strong empirical data to show which subsidiary units have the strong marketing capabilities and why. This is valuable information for the participating MNCs, simply in terms of providing information about where the (perceived) top performers are, but it is also of interest in a more general way because it provides an explanation of why certain units seem to emerge as higher-performers. The managerial conclusion would appear to be that managing the internal systems to enhance communication and absorptive capacity, and remove the secretive culture, are valuable ways of enhancing the marketing capabilities of subsidiary units. However, we prefer to shy away from making such a bold statement until our data collection is complete.

The second important conclusion is that we find reasonable support for the differentiated network model of the MNC. Both external heterogeneity and internal knowledge flows are important factors, the former for creating subsidiary-level differences in the first place, the latter for gradually smoothing out those differences through internal transfers of capabilities. What our results suggest, going back to Figure 8.1, is that we do indeed have to think in terms of the geographical dispersion of marketing capabilities, and the fact that their whereabouts are not well understood, as well as the more practical question of how to facilitate the actual transfer of capabilities between units.

Notes

1. Although we have to show, first of all, that there *is* a significant variation in marketing capabilities in our sample, there is ample previous research that supports this basic premise (Chew *et al.*, 1990; Leibenstein, 1966; Szulanski, 1995).
2. We should also acknowledge an emergent third approach, the 'evolutionary theory of the MNC' (Kogut and Zander, 1995) which focuses on the firm's capacity to transfer tacit knowledge internally as its *raison d'être*.
3. Whether that means that they also exhibit greater variance than other less-dispersed capabilities is an open question (and one we can't address here), but what it should certainly mean is that the identification of leading-edge capabilities is harder than in R&D or manufacturing operations where there are typically many fewer locations to consider.
4. The abbreviations in brackets refer to the labels used in the statistical analysis.

References

Albaum, G., Strandskov, J., Duerr, E. and Dowd, L. (1989) *International Marketing and Export Management*. Reading, MA: Addison-Wesley.

Bartlett, C. and Ghoshal, S. (1986) Tap your subsidiaries for global growth. *Harvard Business Review*, November–December: 87–95.

Bartlett, C. and Ghoshal, S. (1989) *Managing Across Borders: The Transnational Solution*. Boston, MA: Harvard Business School Press.

Birkinshaw, J. M., Hood, N. and Jonsson, S. (1998) Building firm-specific advantage in multinational corporations: the role of subsidiary initiative. *Strategic Management Journal*, 19(3): 221–41.

Birkinshaw, J. M. and Morrison, A. (1995) Configurations of strategy and structure in subsidiaries of multinational corporations. *Journal of International Business Studies*, 26(4): 729–54.

Buckley, P. J. and Casson, M. (1976) *The Future of the Multinational Enterprise*. London: Macmillan.

Cateora, P. R. (1993) *International Marketing*, 8th edn. Homewood, IL: Irwin.

Chew, W. B., Bresnahan, T. F. and Clark, K. B. (1990) Measurement, coordination and learning in a multi-plant network. In R. S. Kaplan (ed.) *Measures for Manufacturing Excellence*. Boston: Harvard Business School Press, pp. 129–62.

Cohen, W. M. and Levinthal, D. A. (1990) Absorptive capacity: a new perspective on learning and innovation. *Administrative Science Quarterly*, 35(1): 128–53.

Cohen, M. D., March, J. G. and Olsen, J. P. (1972) A garbage can model of organizational choice. *Administrative Science Quarterly*, 17: 1–25.

Cyert, R. M. and March, J. G. (1963) *Behavioral Theory of the Firm*. Englewood Cliffs, NJ: Prentice Hall.

Davidson, W. H. (1980) *Experience Effects in International Investment and Technology Transfer*. Ann Arbor: UMI University Press.

Dunning, J. H. (1980) Towards an eclectic theory of international production: some empirical tests. *Journal of International Business Studies*, 11(1): 9–31.

Forsgren, M. and Johanson, J. (eds) (1992) *Managing Networks in International Business*. Philadelphia, PA: Gordon & Breach.

Ghoshal, S. (1986) *The innovative multinational: a differentiated network of organizational roles and management processes*. Unpublished doctoral dissertation, Harvard Business School.

Ghoshal, S. and Bartlett, C. A. (1990) The multinational corporation as an interorganizational network. *Academy of Management Review*, 10(4): 323–37.

Ghoshal, S. and Bartlett, C. A. (1994) Linking organizational context and managerial action: the dimensions of quality of management. *Strategic Management Journal*, 15(special issue): 91–112.

Ghoshal, S. and Nohria, N. (1989) Internal differentiation within multinational corporations. *Strategic Management Journal*, 10(special issue): 323–37.

Håkanson, L. and Nobel, R. (1993). Determinants of foreign R&D in Swedish multinationals. *Research Policy*, 22(5–6): 397–411.

Hedlund, G. (1986) The hyper-modern MNC – a heterarchy? *Human Resource Management*, 25(1): 9–36.

Hedlund, G. (1993) Assumptions of hierarchy and heterarchy, with applications to the management of the multinational corporation. In S. Ghoshal and D. E. Westney (eds) *Organization Theory and the Multinational Corporation*. New York: St Martin's Press.

Hippel, E. von (1994) 'Sticky information' and iteration in the locus of problem solving: implications for innovation. Paper presented at the Prince Bertil Symposium, IIB, Stockholm, June 12–14.

Hofstede, G. (1991) *Cultures and Organizations – Software of the Mind*. Cambridge: Cambridge University Press.

Hymer, S. H. (1976) *The International Operations of National Firms: A Study of Direct Investment*. Boston, MA: MIT Press.

Jarillo, J.-C. and Martinez, J. L. (1990) Different roles for subsidiaries: the case of multinational corporations in Spain. *Strategic Management Journal*, 11(7): 501–12.

Jaworski, B. J. and Kohli, A. K. (1993) Market orientation: antecedents and consequences. *Journal of Marketing*, 57(July): 53–70.

Jeannet, J. P. and Hennessey, H. D. (1995) *Global Marketing Strategies*, 3rd edn. Boston, MA: Houghton Mifflin.

Katz, D. and Kahn, R. L. (1966) *The Social Psychology of Organizations*. New York: Wiley.

Katz, R. and Allen, T. J. (1982) Investigating the not invented here syndrome: a look at the performance, tenure and communication patterns of 50 R&D group projects. *R&D Management*, 12(1): 7–19.

Keegan, W. J. (1989) *Global Marketing Management*, 4th edn. Englewood Cliffs, NJ: Prentice Hall.

Kogut, B. and Singh, H. (1988) The effect of national culture on the choice of entry mode. *Journal of International Business Studies*, 19(3): 411–32.

Kogut, B. and Zander, U. (1995) Knowledge of the firm and evolutionary theory of the multinational corporation. *Journal of International Business Studies*, 26(4): 625–44.

Kohli, A. K. and Jaworski, B. J. (1990) Market orientation: the construct, research propositions and management implications. *Journal of Marketing*, 54(April): 1–18.

Kostova, T. and Cummings, L. L. (1997) Success of the transnational transfer of organizational practices within multinational companies. Paper presented at the Carnegie Bosch Institute's conference, 'Knowledge in International Corporations', Rome, November 6–8.

Kotler, P. (1994) *Marketing Management: Analysis, Planning, Implementation and Control*, 8th edn. Englewood Cliffs, NJ: Prentice Hall.

Kümmerle, W. (1996) *Home base and foreign direct investment in research and development: an investigation into the international allocation of research activity by multinational enterprises*. Unpublished doctoral thesis, Harvard University.

Lawrence, P. and Dyer, D. (1983) *Renewing American Industry*. New York: Free Press.

Lawrence, P. and Lorsch, J. (1967) *Organization and Environment: Managing Differentiation and Integration*. Boston, MA: Harvard University.

Leibenstein, H. (1966) Allocative efficiency *vs* X-efficiency. *American Economic Review*, 56(June): 392–415.

Levitt, T. (1983) The globalization of markets. *Harvard Business Review*, (May–June): 92–102.

Mansfield, E. and Romeo, A. (1980) Technology transfer to overseas subsidiaries by US-based firms. *Quarterly Journal of Economics*, 95: 737–50.

Narver, J. C. and Slater, S. F. (1990) The effect of a market orientation on business profitability. *Journal of Marketing*, (October): 20–35.

Nobel, R. and Birkinshaw, J. M. (1998) Innovation in multinational corporations: control and communication patterns in international R&D operations. *Strategic Management Journal*, 19(5): 479–96.

Nohria, N. and Ghoshal, S. (1997) *The Differentiated Network: Organizing Multinational Corporations for Value Creation*. San Francisco: Jossey-Bass.

Polanyi, M. (1969) *Knowing and Being*. London: Routledge.

Porter, M. E. (1990) *Competitive Advantage of Nations*. Boston, MA: Harvard Business School Press.

Prahalad, C. K. (1976) *The Strategic Process in a Multinational Corporation*. Unpublished doctoral dissertation, School of Business Administration, Harvard University.

Prahalad, C. K. and Doz, Y. L. (1981) An approach to strategic control in MNCs. *Sloan Management Review*, (Summer): 5–13.

Prahalad, C. K. and Doz, Y. L. (1987) *The Multinational Mission*. New York: Free Press.

Quelch, J. A. and Hoff, E. J. (1986) Customizing global marketing. *Harvard Business Review*, (May–June): 2–12.

Rosenzweig, P. and Nohria, N. (1995) Influences on human resource management practices in multinational corporations. *Journal of International Business Studies*, 25(2): 229–52.

Rugman, A. (1981) *Inside the Multinationals: The Economics of Internal Markets*. London: Croom Helm.

Shannon, C. E. and Weaver, W. (1949) *The Mathematical Theory of Communication*. Chicago, IL: University of Illinois Press.

Sharp, B. (1991) Marketing orientation: more than just customer focus. *International Marketing Review*, 8(4): 20–5.

Slater, S. F. and Narver, J. C. (1995) Market orientation and organizational learning. *Journal of Marketing*, 59(July): 63–74.

Sölvell, Ö., Zander, I. and Porter, M. E. (1991). *Advantage Sweden*. Stockholm: Norstedts.

Szulanski, G. (1995) *Appropriating rents from existing knowledge: intra-firm transfer of best practices*. Doctoral dissertation, Paris: INSEAD.

Teece, D. J. (1976) *The Multinational Corporation and the Resource Cost of International Technology Transfer*. Cambridge, MA: Ballinger.

Thompson, J. D. (1967) *Organizations in Action*. New York: McGraw-Hill.

Thompson, J. D. and Tuden, R. (1959) Strategies, structures and processes of organizational decision. In Thompson *et al.* (eds) *Comparative Studies in Administration*. Pittsburgh: University of Pittsburgh Press, pp. 195–216.

Westney, D. E. (1993) Institutionalization theory and the MNC. In S. Ghoshal and D. E. Westney (eds) *Organization Theory and the Multinational Corporation*. New York: St Martin's Press.

White, R. E. and Poynter, T. A. (1990) Organizing for world-wide advantage. In C. A. Bartlett, Y. Doz and G. Hedlund (eds) *Managing the Global Firm*. London and New York: Routledge, pp. 95–113.

Winter, S. G. (1987) Knowledge and competence as strategic assets. In D. Teece (ed.) *The Competitive Challenge – Strategies for Industrial Innovational Renewal*. Cambridge, MA: Ballinger, pp. 159–84.

Woodcock, P. (1994) *The greenfield* vs *acquisition entry mode decision process*. Unpublished doctoral dissertation, Western Business School.

Zander, U. (1991) *Exploiting a technological edge: voluntary and involuntary dissemination of technology*. Doctoral dissertation, Stockholm School of Economics.

Zander, U. and Kogut, B. (1995) Knowledge and the speed of transfer and imitation of organizational capabilities: an empirical test. *Organization Science*, 6(1): 76–92.

9

Knowledge Flows in International Services Firms: A Conceptual Model

Valerie J. Lindsay, Doren Chadee, Jan Mattsson and Robert Johnston

Introduction

Given the increasing involvement of both internal and external network structures and relationships in the internationalization of business (Johanson and Mattsson, 1988), the investigation of services internationalization that involves both parent and foreign subsidiary operations should lead to greater understanding of the dynamics relating to behaviour and performance of these firms (Gupta and Govindarajan, 2000). A key element of these dynamics involves inter- and intra-organizational flows of knowledge. Increased organizational learning and competency development in both parent and subsidiary units of multinational firms requires the effective flow of knowledge between them. In considering the process of knowledge transfer, the role of the individual (Boisot, 1998) also becomes a critical factor to consider. The individual's role in knowledge generation and decision-making in the internationalization process of firms has been largely overlooked until recently, with some insightful contributions being made by Andersson (2000) and Edvardsson *et al.* (1993).

This essay fills a significant gap in the literature on the internationalization of services, by focusing on the process of knowledge flows between parent firms and their foreign subsidiaries, and its associated impact on organizational learning and competency development. We utilize and extend the model of knowledge flows across multinational corporations developed by Gupta and Govindarajan (2000). Unlike these authors, we do not distinguish between knowledge types, but rather, consider the process of knowledge transfer, where knowledge is perceived in a general way. The knowledge transfer process in service firms, however, is more likely to involve context-specific knowledge than in manufacturing firms. While also recognizing important differences between different types of services, for example, in locational embeddedness and context-specificity, this research does not attempt to differentiate between types of service firms.

The role of the individual, as a facilitator in this process, is incorporated into the extended model, drawing mainly on the work of Andersson (2000). The issues of knowledge transfer and individual involvement has particular relevance for service firms, because they tend to rely more on embedded knowledge than manufacturing firms, and people are a key resource. Three specific issues are investigated in the study.

First, the internationalization process is briefly examined from a post-entry perspective, with particular consideration of the foreign market subsidiary, relative to the parent company. Second, the processes relating to knowledge flows and the dyadic relationships between the foreign market subsidiary and the parent are explored. This aims to increase our understanding of the influences of these processes on mutual learning and competency development. Third, the role of the individual in a foreign market subsidiary and/or a parent company in facilitating knowledge flows is investigated. From our understanding of these three issues we develop a conceptual model of the knowledge-based interaction between parent company and foreign market subsidiary in relation to mutual organizational learning and competency development.

The rest of the essay is organized as follows. First, we review the literature on services internationalization, and then consider the role of relationships, knowledge and the individual in this process. The conceptual model developed by Gupta and Govindarajan (2000) is then discussed, along with recent work on the role of individuals in internationalization. These discussions lead to the development of a conceptual model that incorporates aspects of the individual's role in knowledge transfer in the internationalization of services. The conclusion and some research implications, along with the limitations of the study are then presented.

Knowledge flows and internationalization of services

The increase in attention to trade in service industries has led to a large amount of research concerned with the internationalization of services. Most of the research, however, has focused on determining similarities and differences between services and the manufacturing sector (Boddewyn *et al.*, 1986; Dunning, 1989; Vandermere and Chadwick, 1989; Buckley *et al.*, 1992; Chadee and Mattsson, 1998), the internationalization process (Trondsen and Edfelt, 1987; Edvardsson *et al.*, 1993), foreign market entry strategies (Erramilli, 1989, 1991; Erramilli and Rao, 1994), and specific industry or country studies (Sharma, 1989a,b; Johanson and Sharma, 1997). More recently, studies on various behavioural aspects of international services have been undertaken. These include research on cross-cultural differences in service customer satisfaction (Winsted, 1999), global sourcing of services (Kotabe *et al.*, 1998), services innovation (Windrum *et al.*, 1999), services technology (Mattsson, 2000), and communication, both internal and external to the service company (Lievens *et al.*, 1999).

With some exceptions (e.g. Eriksson *et al.*, 1999; Erramilli and Rao, 1994; Patterson and Cicic, 1995), there has been limited contribution towards the development of theory and constructs that are able to generalize across service industries (Knight, 1999). Because the international services domain is so complex and diverse, it may not be possible to develop valid theories. We believe that a more eclectic approach to understanding internationalization should attempt to model the critical processes of knowledge transfer that help to explain the motivation of firms to internationalize, their foreign market entry decision and further foreign market development. For example, O'Farrell *et al.* (1998), highlighting the limitations of existing theories in explaining international market development by service firms, suggest the need for a more behavioural approach. They propose a strategic choice framework which encompasses the strategic complexities that arise when new opportunities need to be balanced by higher risks of international market development. In a similar way, Edvardsson *et al.* (1993) propose a dynamic, creative, and behavioural perspective, rather than a planning perspective. Key to this is the role of the entrepreneur, particularly in the early stages of internationalization, when information and knowledge gained through the entrepreneur's networks are translated into strategy development.

Samiee (1999) identifies three key issues for services marketing. First, services will play a more important role in the marketing of goods as firms endeavour to seek competitive advantage through their value chains, which are becoming increasingly complex. Second, goods and services are increasingly less distinguishable, with every tangible product containing some service (Lindsay, 1999). Third, significant issues arise from the classification of services. Most classification systems incorporate a number of fundamental features of services, the most important of which are the separability of the service and its spatial features. The degree of separability of the service from its associated physical good is a major influence on the deployment of people in the marketplace and thus the market entry strategy of the firm (Boddewyn *et al.*, 1986; Vandermere and Chadwick, 1989; Buckley *et al.*, 1992). For example, for a service with a high degree of inseparability (production cannot be separated from consumption), a foreign market presence is necessary (Boddewyn *et al.*, 1986). Also the degree to which production and consumption of service can be spatially separated will influence the location of activities abroad (Buckley *et al.*, 1992). The locational and physical inseparability of services suggest that services-related knowledge is high in context-specificity. The subtleties of context can generally only be shared by personal (usually face-to-face) interaction between individuals, in a process sometimes known as 'handshaking' (Leamer and Storper, 2001). These factors indicate the need for close relationships and effective knowledge flows between the foreign market operation of the service firm and its parent company, in order to maintain and support the firm's strategy in its foreign market. The many dimensions, which differentiate services from goods highlight the strategic importance of knowledge generation and knowledge transfer, and the strong relational characteristics associated with the internationalization of services (O'Farrell, 1998).

Relatively little is known about the influence of national culture on service provision, partly because the area lacks depth in empirical studies (Mattsson, 2000). Services entail essentially social encounters, and these are likely to vary across different cultures (Winsted, 1999). A foreign market culture may have a significant impact on the acceptability and adoption pattern of services (Samiee, 1999). The higher involvement of people in service industries than in manufacturing stresses the importance of culture, and potentially exposes the firm to higher levels of cultural incompatibility. In addition, since the requirement for local adaptation is generally high, standardization is less likely to occur with services than with products. All of these factors suggest that knowledge and management of cultural differences is important for international service firms. The strategic choice framework developed by O'Farrell *et al.* (1998) highlights the need for firm-related cultural expertise in responding to expanding foreign demand, such as specific language skills, international strategic know-how and cultural and regulatory experience. Given these considerations, effective information and knowledge flows into and out of the firm's foreign market operations, and between them and their parent companies are critical. Country culture has been shown to impact on knowledge transfer, with cultural similarity generally supporting the transfer of knowledge in a largely undistorted form (Bhagat *et al.*, 2002). In addition, absorptive capacity can be influenced by cultural elements in a country (Kedia and Bhagat, 1988).

Most service related research to date has focused largely on pre-entry processes of internationalization (Wind *et al.*, 1973), market entry strategies, and studies of specific behavioural characteristics in the foreign market (Knight, 1999). The perspective taken has generally been that of the originating service firm (i.e. parent company), rather than the firm's foreign market actors, or subsidiaries. However, there is increasing recognition of the need for post-market entry research, which considers the perspective of the foreign market actors, as well as the home-based unit of the company (O'Farrell *et al.*, 1998), as a larger internal network. Maintenance of an effective internal network relies on good communication and knowledge flows between various players in the organization – in the case of multinational companies, between the parent company and its subsidiaries.

Relationships as a key element of knowledge flows

Knowledge transfer occurs in a shared social context between linked units (Huber, 1991; Tsai, 2001). Relationships between various actors in international services represent an important determinant of knowledge transfer (Windrum and Tomlinson, 1999), competency development and perceived service quality (Eriksson *et al.*, 1999). Service quality is created in relationships between suppliers and customers (Holmlund and Kock, 1995; Majkgård and Sharma, 1998), where both are active participants in the process (Håkansson, 1982). The exchange of information allows each party to determine the other's requirements and develop perceptions of service quality (Eriksson *et al.*, 1999). Relationships also enable the sharing of resources, which may involve technical or functional aspects of knowledge, capital, products and other services. Through these processes the service supplier and service buyer adapt to each other (Eriksson *et al.*, 1999). When a service is standardized, or where the foreign market operation is reliant on competences and information from the parent company in the provision of the service to their customers, effective relationships between the parent and foreign market operation are crucial in facilitating the necessary information and knowledge flows (Buckley and Carter, 1999).

Both internal and external relationships are key prerequisites for successful internationalization in service firms (Edvardsson *et al.*, 1993). External relationships are of two types: vertical and horizontal (Mattsson, 1985); these can be either domestic or international, or both. Vertical relationships (often called customer, or client, relationships) are those between a seller and his/her customers, and horizontal relationships (often called industry relationships) are those between a firm and other firms in the same industry. Key elements of business-to-business relationships are trust, commitment and adaptation (Hallen *et al.*, 1991; Morgan and Hunt, 1994). External relationships incorporate Mattsson's (1985) vertical and horizontal relationships, including those with customers and other service suppliers, particularly of supplementary services (Kotabe *et al.*, 1998), and other companies in the same industry (Edvardsson *et al.*, 1993). Internal relationships between parts of the organization enable the dynamic interplay between the functional elements of the organization involved in international services. International businesses are usually involved in widely spread internal and external networks, and it has been suggested that relational studies should try to encompass these wider links (Gupta and Govindarajan, 2000).

Knowledge: an integral element in services

Knowledge exists in relationships between individuals (Boisot, 1998; De Long and Fahey, 2000), and relationships facilitate the flow of information. Information is a critical factor in the production of many services, and competitive advantage is likely to accrue to service firms that have differential access to information (Buckley *et al.*, 1992). This requires capabilities in gathering, storing, monitoring, exhibiting and analysing information at the lowest possible cost. Information, experience and knowledge are closely linked since knowledge can be derived from investment in information, or from the experience of key personnel. Boisot (1998) suggests that information becomes knowledge when it is accompanied by a propensity towards activity.

In studying knowledge-intensive services (KIS), Windrum and Tomlinson (1999) note how the innovativeness of these firms can contribute to national innovation, through their interaction with non-service sectors. Knowledge-intensive service firms contain a high degree of tacit knowledge (Starmbach, 1997). Through interaction with their clients, knowledge-intensive firms not only transfer knowledge, but also engage in the co-production of knowledge with the client (Windrum and Tomlinson, 1999). The quality of this interaction depends on the competences of the client as well as those of the knowledge-intensive service firm (O'Farrell, 1998). Organizational routines (Levitt and March, 1988; Cohen and Levinthal, 1990) and culture, including practices, beliefs and values (Ciborra, 1993) have also been associated with the adoption and absorption of new knowledge.

In recent years, research has also turned to the firm's internal management of knowledge and the effectiveness of knowledge utilization from the transfer process between parent and subsidiary (e.g. Buckley and Carter, 1999). In their research on multinational corporations, Gupta and Govindarajan (2000) studied the flows of knowledge to and from subsidiaries, and considered the factors that influence the inflow and outflow of knowledge from these operational units. While their study only considered the nodal unit, for the sake of simplicity, Gupta and Govindarajan (2000) conceptualize knowledge flows in a way that can be applied to dyadic and wider network arrangements.

Based on the elements recognized in communication theory, Gupta and Govindarajan (2000) conceptualized knowledge flows to be a function of the following five factors: (1) the value of the source unit's knowledge stock – that is, the value of a subsidiary's knowledge stock to other units in the company's global network; (2) motivational disposition of the source units – that is, a subsidiary's disposition to overcome tendencies to enjoy an 'information monopoly', and share and receive information with other units openly; (3) existence and richness of transmission channels – that is, transmission channels characterized by openness, informality and density of communications; (4) motivational disposition of the target unit – as with the source unit, the disposition of the parent firm to share and receive knowledge, and overcome tendencies, such as the 'not invented here' syndrome; (5) absorptive capacity of the target unit – that is, a firm's ability to recognize the value of new information, assimilate it and apply it to commercial ends (Cohen and Levinthal, 1990).

Although the conceptual framework of Gupta and Govindarajan (2000) provides a useful lens through which to view knowledge transfer, it is limited to the extent that it considers knowledge flows to and from subsidiaries only. In this essay, we build on the conceptual framework of Gupta and Govindarajan to develop a broader framework, applicable to international service firms, where either the parent company or the subsidiary units can be the source of, or target for, knowledge. In particular, we consider the perspective of the subsidiary (or foreign market operation) and suggest that knowledge transfer from the parent to the subsidiary is strategically important. While traditional views of the MNC emphasize the parent, or headquarters, as the central knowledge-capturing and exploiting part of the MNC, recent network interpretations of the MNC support a more systemic approach (e.g. Bartlett and Ghoshal, 1989). Foss and Pedersen (2002) suggest that knowledge flows are influenced by the knowledge structure of the MNC. For example, in the differentiated MNC, subsidiary knowledge is distant from, or peripheral to, the core of the knowledge structure. In this situation, the subsidiary is more likely to source external knowledge, and the parent is less likely to share internal knowledge.

An advantage of this structure is the resulting heterogeneity among subsidiaries and parent, but a disadvantage is the limited access to company-wide knowledge, unless the parent adopts a deliberate coordination role (Bartlett and Ghoshal, 1989). We concur with others that the MNC network structure enables subsidiaries to access relevant strategic resources (Tsai, 2001), and that the headquarters plays a key role in coordinating the process of knowledge transfer and resource-sharing (Gresov and Stephens, 1993). The coordination process involves the transfer of knowledge at a systemic level in the MNC, and includes the headquarters, or parent, as a direct knowledge source.

The individual: main driver of knowledge flows

While knowledge is a firm's resource based competency (Grant, 1996), the dynamics underlying the relationship between knowledge and the firm have not been well established. The firm is, essentially, a repository of knowledge, and ownership of knowledge resides in individuals within the firm (Arthur *et al.*, 1995; Boisot, 1998). This is particularly apparent when knowledge is of a tacit or uncoded nature (Nonaka and Takeuchi, 1995). Since individuals may move from firm to firm during their careers, the firm has no guaranteed long-term access to, or utility from, their unique, tacit knowledge.

In research on inter- and intra-firm relations, the flow of knowledge is often conceptualized as a firm level property (Gupta and Govindarajan, 2000). Individuals are considered mainly in the role of the facilitating entrepreneur or firm leader or CEO in the process of knowledge transfer between firms (Nonaka, 1991). However, it is also acknowledged that individuals throughout the firm play a significant role in the process of knowledge accumulation, flow and utilization (Nonaka, 1991). In the MNC context, the individual is a key actor in the transfer of knowledge between parent company and foreign subsidiary operations. This has been noted in other studies, particularly across different industries (Lindsay *et al.*, 2001).

Given the importance of the individual in the process of knowledge transfer, it is somewhat surprising that the role of the individual in international business was largely neglected until the 1990s (Andersson, 2000). While Andersson focuses on the entrepreneur, we do not specifically characterize the individual as the entrepreneur. Rather, we refer to the 'individual' in a wider sense, to denote any person in an organization, whether or not they are 'entrepreneurial'. Edvardsson *et al.* (1993) suggest that internationalization processes are driven by individual actors and entrepreneurs, and are characterized by the extent to which these people think in an international perspective. Similarly, Andersson (2000) considers internationalization as an integral part of the firm's strategy, with the entrepreneur having the main responsibility of linking the process (strategy) with the structure (environments), through the transfer of relevant information and knowledge. Thus, in Andersson's model, the entrepreneur is central to decision-making about international market choice and entry mode. Andersson (2000) recognizes three types of entrepreneur, depending on his/her behavioural context, skill base and interpretation of the environments in which (s)he operates. These are marketing, structural and technology entrepreneurial types. The different entrepreneurial approaches lead to the emergence of different types of internationalization strategy.

Successful internationalization of the service firm is dependent on information and its interpretation as knowledge by individuals. Knowledge about the environments influencing the subsidiary can be enhanced, or inhibited by, the flow of knowledge between individuals in the parent company and subsidiary, particularly through their dyadic relationships. In order to facilitate effective knowledge flows, the individual in a subsidiary must maintain complex relationships with its parent company, other subsidiaries, and its usual suppliers and customers.

The distinctions in the literature between individual and organizational knowledge and learning are inconclusive (Kim, 1993; Argyris, 1999), with a few notable exceptions (e.g. Nonaka and Takeuchi, 1995). A specific role for the individual is seldom included in the conceptualization and operationalization of organizational learning. There is general agreement, however, that organizational learning results from the embeddedness of individual knowledge in the organization, for example, in competencies, relationships, organizational culture (Argyris, 1999), rules (March *et al.*, 2000), and routines (Levitt and March, 1988). Hult (1998) suggests that the MNC integrates and transfers knowledge to other members in the MNC network through the organizational learning structure, indicating the importance of knowledge sharing with its subsidiaries by the parent company.

Knowledge flows between parent and subsidiary in service firms

The following discussion describes the conceptual model, drawing on the theoretical perspectives outlined above. Key components in the model are identified in the discussion in italics. The conceptual model of knowledge transfer developed in Figure 9.1 draws directly from two

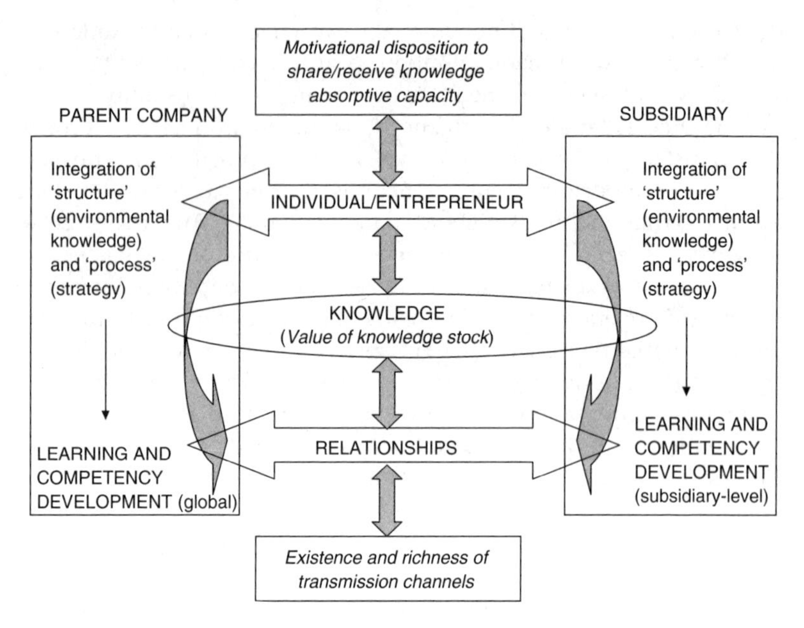

Figure 9.1 Conceptual model of knowledge flows between parent and subsidiary

sources: Gupta and Govindarajan's (2000) model of knowledge flows within multinational corporations, and Andersson's (2000) model of the entrepreneur in the internationalization process. Both of these models have been discussed in earlier sections in this essay. In addition, key elements discussed in the literature review complement the other components in the conceptual model.

Our model depicts the process of knowledge transfer between parent and foreign subsidiaries. We use the five main components of Gupta and Govindarajan's (2000) model to depict the knowledge transfer process, and also incorporate other components, including adaptations from Andersson's (2000) model. In contrast to Gupta and Govindarajan, our model considers *knowledge flows* between parents and foreign operations, rather than solely into, and out of, the subsidiary. In other words, we take a dyadic, rather than nodal, perspective. A further point of difference is the explicit role of *relationships* in the conceptual model, highlighted because they play a significant part in the knowledge transfer process, particularly in service firms.

Our model also relates knowledge flows to *learning and competency development* in both parents and subsidiary firms, encompassing elements from Andersson's (2000) model. Specifically, the conceptual model considers the *integration of environmental knowledge (structure)* with *strategy (process)* in the parent and subsidiary companies, and the impact of these elements on knowledge flows. The *individual* is central to the process of knowledge transfer, being the creator and transmitter of knowledge (Nonaka, 1991; Boisot, 1998). Individuals embody the behavioural elements associated with knowledge and knowledge transfer. In the conceptual model, individuals link the parent firm and its foreign subsidiary through dyadic relationships. They create effective *knowledge transmission channels* through relationships that are trusting, open and informal (Rogers, 1995). Individual knowledge is translated into organizational learning and competency development particularly through the conversion of tacit to explicit knowledge (Nonaka, 1991).

The *motivational disposition of the parent and subsidiary to share and receive knowledge* is fundamentally a property of individuals. The personalities of individuals may influence the ease with which knowledge transfer occurs. Often, negative personality traits emerge, such as ego defence mechanisms, jealousies, territorial protection etc. Similarly *absorptive capacity* depends on the extent to which values and characteristics are shared between individuals in the parent company and subsidiary operation. Shared knowledge about personal characteristics facilitates the development of relationships between the actors concerned. Relationships and associated communication processes between two firms both influence, and are influenced by, the existence and richness of the knowledge transmission channels.

The *value of the knowledge stock* is a perceived quality, again, a predominantly individual characteristic. Perceptions may vary between individuals in the parent and subsidiary companies, and may be influenced by the quality of their relationship, and shared values etc. The value of the knowledge stock relates to the parent company strategy, particularly with regard to its degree of globalization and cross subsidiary learning, and the potential value of the knowledge to the MNC at a global level.

The following discussion illustrates how the components of the model are integrated. As shown in the centre of the model (Figure 9.1), individuals in the subsidiary and the parent company create and transfer knowledge through their dyadic relationships. The effectiveness of the knowledge transfer depends on a number of factors. First, the quality of the relationship, particularly in terms of trust and shared socialization characteristics, influences the extent to which knowledge is shared. The better the relationship in these dimensions, the more effective the transmission channels and the greater the degree of knowledge shared.

Second, the motivation of each organization to share and receive knowledge, and their absorptive capacity, influences the extent of knowledge transfer. If, for example, the parent company is poorly motivated, the subsidiary will lack parental, and probably corporate, knowledge and be limited in its ability to share its own knowledge with the parent. The parent's propensity to receive knowledge from the subsidiary may also be influenced by its perceived value of the knowledge, particularly in terms of its global utility. Third, each side of the model represents the integration of knowledge (both internal and external) gained by the subsidiary or parent with their strategy, in accordance with Andersson's (2000) model. Important outcomes from these knowledge-based inputs are organizational learning and competency development, which are critical to a firm's competitiveness (Hamel and Prahalad, 1994), and reflect the collective knowledge capital of individuals in the firm.

In summary, the conceptual model attempts to cast light on the mechanisms underpinning the process of knowledge transfer between parent company and subsidiary in international service firms. Both Gupta and Govindarajan's (2000) knowledge transfer model and Andersson's (2000) entrepreneurship model provide a helpful opportunity to explore this issue.

Discussion and implications

The conceptual model developed in this essay aims to create a better understanding of the mechanisms underpinning the process of knowledge transfer in service multinationals. The model extends that of Gupta and Govindarajan (2000) in two main ways. First, it incorporates the role of the individual in facilitating knowledge flows, particularly through relationship building. Second, the model considers the dyadic relationship and associated knowledge flows between foreign subsidiary operations and their parent firms, rather than just the subsidiary, or nodal unit. A number of key points emerge from the conceptual model, discussed as follows.

The individual is a central component, through which knowledge is created and transferred. Relationships between individuals within and outside the organization (the parent and foreign operation) facilitate this process. The model captures the translation of individual knowledge to organizational learning and competency development through the integration of knowledge with the firm's strategy. Specifically, individual knowledge becomes embedded in the organization in the process of organizational learning, and so the organization is less exposed to the career mobility (Arthur *et al.*, 1995) of its employees. Embedded knowledge is evident in, or example, organizational competencies, routines, and standard practices (Argyris, 1999), and, ultimately, in a firm's products, services and processes (Demarest, 1997). In facilitating organizational learning, the firm must effectively convert individual tacit knowledge into explicit knowledge, which has utility at the organizational level.

The model presents some unique challenges for service firms. A potential paradox exists, for example, with the component, value (to the parent company) of source unit's (foreign market subsidiary) knowledge stock, in the following way. Adaptation and customization in foreign markets is generally more important in the service sector than the manufacturing sector (Edvardsson *et al.*, 1993). According to Gupta and Govindarajan (2000) the parent company values information and knowledge from its subsidiaries when these can be shared strategically amongst other subsidiaries, for the benefit of the whole organization. However, when local markets demand specific adaptation of services, and often customization to individual customer needs, the resulting services are less likely to benefit other subsidiaries. In this case, the parent company will likely place a low value on the subsidiary's information and knowledge, and be less inclined to participate fully in the knowledge transfer process. Even though the parent company itself might possess valuable and relevant knowledge, it will contribute little to the adaptive responses of the subsidiary. This, in turn, impacts on the ability of the subsidiary to service its local market and maintain close customer relationships.

The situation may also be explained by differences in knowledge types. To have high value, a subsidiary's knowledge stock must be generalizable across different markets. 'Knowing how' types of knowledge, that is knowledge of a more tacit (Nonaka, 1991), or uncoded (Boisot, 1998) nature, will be more valuable in this regard, than 'knowing what' types of knowledge, that are more informational in nature (Gupta and Govindarajan, 2000). The management of knowledge transfer is, therefore, more complex when different types of knowledge are considered.

The motivational disposition of the parent company to share and receive knowledge is strongly influenced by its perception of the value of the knowledge. It is not unusual for the motivations of the parent company and the foreign operation to differ. For example, the foreign market operation may be highly motivated to share and receive knowledge, while the parent company may have little inclination to do so. This may result from differences in perceived value of the knowledge, or be an ego-defence mechanism (Allport, 1937). Conversely, the foreign market operation may chose to retain its knowledge, even though it may be of value to the parent company.

Communication channels between the parent and foreign operation act as transmission channels for knowledge transfer between the two parties. Channel effectiveness depends largely on the perceived value and motivation to receive and share knowledge. The existence and richness of these channels are driven largely by the quality of the relationships between individuals in the parent company and foreign operation. Relationships characterized by informality and openness appear to enrich the transmission channels.

The absorptive capacity of the target units, that is both parent company and foreign operations as recipients of knowledge, is also deeply rooted in individual relationships. These influence the extent of inter-unit as well as sharing of common meanings, and personal and social characteristics, which enhance communication and have greater effects in terms of knowledge gain (Rogers, 1995). Variations in knowledge transfer between the case companies in a recent study of services internationalization (Lindsay *et al.*, 2001) appear to result mainly from differences in the quality of relationships and inter-unit homophily.

Given the developing interest in the literature in the role of knowledge in firm competitiveness, the development of conceptual frameworks for understanding the dynamics associated with this process is important. This is especially important for service firms, which tend to rely much more intensely on knowledge and its transfer than manufacturing firms, for reasons outlined earlier. While the conceptual model presented in this essay undoubtedly requires refinement and empirical validation, we believe that it brings together some of the key elements addressing the issue of knowledge transfer across international service firms.

The implications for further research touch on three main areas. First, there is a need for further insights into the behavioural aspects associated with services internationalization, particularly on the role of knowledge transfer, which is fundamental to the operationalization of services. Second, the dual relationships between services-based parent firms and their foreign subsidiaries, in terms of knowledge transfer and the resultant learning and competence development, needs considerably more investigation. Third, there is a need to generate a better understanding of the influence and role of the individual in relation to the above two contexts. Contemporary views of the firm and the increasing mobility of human capital, demand a greater focus on the individual in relation to firm competitiveness.

Managers of firms involved in international service industries would benefit from a clearer understanding of these three aspects. Understanding the links between the components of our conceptual model should enable managers to determine how knowledge flows between parent and subsidiary operations can be both promoted and inhibited. The model exposes some key organizational issues associated with its separate components and their integration, such as employee recruitment, motivation and reward systems, training for innovation, and other strategic knowledge issues, such as the conversion of tacit to explicit knowledge.

The conceptual model has been developed from prior empirical studies of service industries (Lindsay *et al.*, 2001). It is not possible, at this stage, to ascertain the generalizability of the model, such as the extent to which the underlying concepts are applicable to other settings involving knowledge flows across dyadic relationships between companies. Testing of the model across a variety of settings could be an early follow-on from this work. Although some causal possibilities between components of the model were identified in this paper (e.g. between individuals, relationships and knowledge flows), these have not been established empirically. Research determining the relationships between different components of the model would contribute valuable insights into the knowledge transfer process between parent companies and their subsidiary operations.

The study has not specifically operationalized the components of the model, but has developed a qualitative perspective on events and activities in international service firms from an earlier case study basis. Each of the components now need to be examined in more detail, both in their own right, and as an integrated whole, as depicted in the model. Since relatively little research has been conducted in the area of knowledge flows in international service firms, it is important to encompass concepts and ideas established in the general literature on knowledge transfer, in order to develop a better understanding of the knowledge dynamics of services internationalization.

References

Allport, G. W. (1937) *Personality: A Psychological Interpretation*. New York: Holt.

Andersson, S. (2000) The internationalization of the firm from an entrepreneurial perspective. *International Studies of Management and Organization*, 30(1): 63–92.

Argyris, C. (1999) *On organizational Learning*, Oxford: Blackwell.

Arthur, M. B., Claman, P. H. and De Fillippi, R. J. (1995) Intelligent enterprise, intelligent careers. *Academy of Management* Executive, 9(4): 7–20.

Bartlett, C. A. and Ghoshal, S. (1989) *Managing Across Borders: The Transnational Solution*. Boston: Harvard Business School Press.

Bhagat, R. S., Ben, B. L., Harveston, P. D. and Triandis, H. C. (2002) Cultural variations in the cross-border transfer of organizational knowledge: an integrative framework. *Academy of Management Review*, 27(20): 204–22.

Boddewyn, J. J., Halbrich, M. B. and Perry, A. C. (1986) Service multinationals: conceptualization, measurement and theory. *Journal of International Business Studies*, 16(3): 41–57.

Boisot, M. H. (1998) *Knowledge Assets: Securing Competitive Advantage in the Information Economy*. Oxford: Oxford University Press.

Buckley, P. J. and Carter, M. J. (1999) Managing cross-border complementary knowledge. *International Studies of Management and Organization*, 29(1): 80–104.

Buckley, P. J., Pass, C. L. and Prescott, K. (1992) The internationalization of service firms: a comparison with the manufacturing sector. *Scandinavian International Business Review*, 1(1): 39–56.

Chadee, D. and Mattsson, J. (1998) Do service and merchandise exporters behave and perform differently? A New Zealand investigation. *European Journal of Marketing*, 32(9/10): 830–42.

Ciborra, C. U. (1993) *Teams, Markets, Systems: Business Innovation and Information Technology*. Cambridge: Cambridge University Press.

Clark, T., Rajaratnum, D. and Smith, T. (1996). Towards a theory of international services: marketing intangibles in a world of nations, *Journal of International Marketing*, 4(2): 9–28.

Cohen, W. M. and Levinthal, D. A. (1990) Absorptive capacity: a new perspective on learning and innovation. *Administrative Science Quarterly*, 35:128–52.

De Long, D. W. and Fahey, L. (2000) Diagnosing cultural barriers to knowledge management. *Academy of Management Executive*, 14 (4): 113–28.

Demarest, M. (1997) Understanding knowledge management. *Long Range Planning*, 30(3): 374–84.

Dunning, J. H. (1989) Multinational enterprises and the growth of services: some conceptual and theoretical issues. *Service Industries Journal*, 9(1): 5–39.

Edvardsson, B., Edvinsson, L. and Nystrom, H. (1993) Internationalization in service companies. *Services Industries Journal*, 13(1): 80–97.

Eriksson, K., Majkgård, A. and Sharma, D. D. (1999) Service quality by relationships in the international market. *Journal of Services Marketing*, 13(4/5): 361–75.

Erramilli, M. K. (1989) Entry mode choice in service industries. *International Marketing Review*, 7(5): 50–62.

Erramilli, M. K. (1991) The experience factor in foreign market entry behaviour of service firms. *Journal of International Business Studies*, 7: 479–501.

Erramilli, M. K. and Rao, C. P. (1994) Service firms' international entry-mode choice: a modified transaction-cost analysis approach. *Journal of Marketing*, 57: 19–38.

Foss, N. and Pedersen, T. (2002) The MNC as a knowledge structure: the role of knowledge sources and organizational instruments for knowledge creation and transfer. Working paper, Copenhagen Business School, Denmark.

Grant, R. M. (1996) Towards a knowledge-based view of the firm. *Strategic Management Journal*, 17: 109–22.

Gresov, C. and Stephens, C. (1993) The context of interunit influence attempts. *Administrative Science Quarterly*, 38: 252–76.

Gupta, A. K. and Govindarajan, V. (2000) Knowledge flows within multinational corporations. *Strategic Management Journal*, 21(4): 473–96.

Håkansson, H. (1982) *International Marketing and Purchasing of Industrial Goods. An Interaction Approach*. Chichester: John Wiley.

Hallen, L., Johanson, J. and Mohamed, N.-S. (1991) Interfirm adaptation in business relationships. *Journal of Marketing*, 55(2): 29–37.

Hamel, G. and Prahalad, C. K. (1994) *Competing for the Future*. Boston, MA: Harvard Business Press.

Holmlund, M. and Kock, S. (1995) Buyer perceived service quality in industrial networks. *Industrial Marketing Management*, 24: 109–21.

Huber, G. (1991) Organizational learning: the contributing processes and the literatures. *Organization Science*, 2(1): 88–115.

Hult, G. T. M. (1998) Managing the international strategic sourcing process as a market-driven organizational learning system. *Decision Sciences*, 29(1): 193–216.

Johanson, J. and Mattsson, L.-G. (1988) Internationalization in industrial systems: a network approach. In N. Hood and J.-E. Vahlne (eds) *Strategies in Global Competition*, New York: Croom Helm.

Johanson, J. and Sharma, D. D. (1997) Technical consultancy in internationalization, *International Marketing Review*, 4(Winter): 20–9.

Kedia, B. and Bhagat, R. S. (1988) Cultural considerations on transfer of technology across nations: implications for research in international and comparative management. *Academy of Management Review*, 13(4): 559–71.

Kim, D. H. (1993) The link between individual and organizational learning. *Sloan Management Review*, 35(4): 37–50.

Knight, G. (1999) International services marketing: review of research, 1980–1998. *Journal of Services Marketing*, 13(4/5): 347–60.

Kotabe, M., Murray, J. Y. and Javalgi, R. G. (1998) Global sourcing of services and market performance: an empirical investigation. *Journal of International Marketing*, 6(4):10–31.

Leamer, E. E. and Storper, M. (2001) The economic geography of the Internet age. *Journal of International Business Studies*, 32(4): 641–65.

Levitt, B. and March, J. G. (1988) Organizational learning. In W. R. Scott (ed). *Annual Review of Sociology*, 14: 319–40.

Lievens, A., Moenaert, R. K. and S'Jegers, R. (1999) Linking communication to innovation success in the financial services industry: a case study analysis. *International Journal of Service Industry Management*, 10(1): 23–47.

Lindsay, V. J. (1999) *A Strategic View of Export Performance: A New Zealand Perspective*. PhD, Marketing and Strategic Management Group, University of Warwick, Warwick.

Lindsay, V. J., Johnston, R., Mattsson, J., Chadee, D. and Millett, B. (2001) Relationships, the role of individuals and knowledge flows in the internationalization of service firms. Working paper, Department of International Business, the University of Auckland, New Zealand.

Majkgård, A. and Sharma, D. D. (1998) Service quality by international relationships: service firms in the global market. In C. P. Rao (ed.) *Globalization, Privatization and the Free Market Economy*, New York, NY and Westport, CT: Quorum Books.

March, J. G., Schulz, M. and Zhou, X. (2000) *The Dynamics of Rules: Studies of Change in Written Organizational Codes*. Stanford, CA: Stanford University Press.

Mattsson, J. (2000) How to manage technology during services internationalization. *Service Industry Journal*, 20(1): 22–39.

Mattsson, L. -G. (1985) An application of a network approach to marketing: defending and changing market positions. In N. Dholakia and J. Arndt (eds) *Alternative Paradigms for Widening Market Theory*, Greenwich, CT: JAI Press.

Morgan, M. R. and Hunt, S. D. (1994) The commitment-trust theory of relationship marketing. *Journal of Marketing*, 58: 20–38.

Nonaka, I. (1991) The knowledge-creating company. *Harvard Business Review*, 91(November/December): 312–20

Nonaka, I. and Takeuchi, H. (1995) *The Knowledge-Creating Company*. Oxford: Oxford University Press.

O'Farrell, P. N., Wood, P. A. and Zheng, J. (1998) Regional influences on foreign market development by business service companies: elements of a strategic context explanation. *Regional Studies*, 32(1): 31–48.

Patterson, P. G. and Cicic, M. (1995) A typology of service firms in international markets: an empirical investigation. *Journal of International Marketing*, 3(4): 57–84.

Rogers, E. M. (1995) *Diffusion of Innovations*. New York: Free Press.

Samiee, S. (1999) The internationalization of services: trends, obstacles and issues. *Journal of Services Marketing*, 13(4/5): 319–28.

Sharma, D. D. 1989(a) Overseas market entry strategy: the technical consultancy firms. *Journal of Global Marketing*, 2(2): 89–110.

Sharma, D. D. 1989(b) *Technical Consultancy as a Network of Relationships: Advances in International Marketing*. Greenwich, CT: JAI Press.

Starmbach, S. (1997) *Knowledge-Intensive Services and Innovation in Germany*. Stuttgart: University of Stuttgart.

Trondsen, E. and Edfelt, R. (1987) New opportunities in global services. *Long Range Planning*, 20(5): 53–61.

Tsai, W. (2001) Knowledge transfer in intraorganizational networks: effects of network position and absorptive capacity on business unit innovation and performance. *Academy of Management Journal*, 44(5): 996–1005.

Vandermere, S. and Chadwick, M. (1989) The internationalization of services. *Service Industries Journal*, 9(1): 79–93.

Wind, Y., Douglas, S. and Perlmutter, H. V. (1973) Guidelines for developing international marketing strategies. *Journal of Marketing*, 37(April): 14–23.

Windrum, P. and Tomlinson, M. (1999) Knowledge-intensive services and international competitiveness: a four-country comparison. *Technology Analysis and Strategic Management*, 11(3): 391–408.

Winsted, K. F. (1999) Evaluating service encounters: A cross-cultural and cross-industry exploration. *Journal of Marketing Theory and Practice*, 7(2): 106–23.

10
The Dilemmas of MNC Subsidiary Transfer of Knowledge

Jens Gammelgaard, Ulf Holm and Torben Pedersen

Introduction

To an increasing extent, the success of multinational companies (MNCs) is considered to be contingent upon the ease and speed by which valuable knowledge is disseminated throughout the organization (Hedlund, 1986; Bartlett and Ghoshal, 1989; Gupta and Govindarajan, 1991). Thus, creation of knowledge in a spatially dispersed multinational organization and tapping into advanced local knowledge wherever it can be found are necessary conditions for success in the global marketplace. The implication is that some subsidiaries are supposed to act as bridgeheads (Forsgren *et al.*, 1999) that tap into knowledge created in a local context and subsequently transfer the knowledge to other MNC units where it is of better use. Therefore, some subsidiaries will or ought to have a strategic role in the global organization that reaches beyond their local undertakings (e.g. Gupta and Govindarajan, 1994; Holm and Pedersen, 2000). However, there are obstacles to the internal transfer of knowledge in the MNC, and a number of dilemmas unfold within those subsidiaries that are supposed to ensure internal knowledge transfer.

The obstacles to knowledge transfer are manifold (e.g. Szulanski, 1996; Gupta and Govindarajan, 2000), and relate to both cognitive factors (e.g. tacitness, ambiguity and context-specificity) and issues related to either the sender or receiver's motivation. Knowledge transfer is far from smooth, but rather demanding in terms of the time and resources necessary to overcome these obstacles. For the MNC subsidiary, this creates a number of dilemmas. Should knowledge be acquired externally or internally? To what extent should the limited resources be spent on external knowledge acquisition versus internal knowledge transfer? In this chapter, obstacles and dilemmas related to internal knowledge transfer are discussed in terms of a unique empirical data set on subsidiaries and their internal and external knowledge linkages. We reveal empirical findings from the Centres of Excellence project and discuss how subsidiaries handle dilemmas related to the transfer of knowledge.

Different types of dilemma

Research on knowledge flows has attempted to identify some factors that inhibit or facilitate knowledge flows between MNC units. Szulanski (1996) explored 'internal stickiness' of knowledge, i.e. factors that impede the intra-firm transfer of knowledge. He found that transferring knowledge within a corporation is far from easy. He identified two sets of factors, motivational and knowledge-related, which impede the internal knowledge transfer (internal stickiness). The latter stems from the tacit, context-specific and ambiguous kind of knowledge, which is difficult to transfer from one location to another, while the former is related to the subsidiary's motivation to apply the time and resources necessary to conduct the transfer. The knowledge-related impediments to knowledge transfer stem from the observation that knowledge development is context or relation specific. Knowledge is most valuable in its own context, while there may be obstacles to applying the knowledge in a different context. Therefore, the knowledge may not be useful for other MNC units operating in other local environments. In essence, there is an inherent conflict between a subsidiary's ability to create new knowledge, on one hand, and the possibility to transfer and use that knowledge within the MNC on the other (Forsgren *et al.*, 2000).

Problems arise when knowledge is moved from one context to another, and the receiver loses some, if not all, of the knowledge's original meaning. For an outsider, such as a company's headquarters, to retrieve local embedded knowledge, such as knowledge generated in the relationship between a subsidiary and a specific customer, is often difficult and sometimes impossible. Even if the knowledge is codified, articulated and stored in the organizational memory, a substantial distance exists between the informant who encodes data and the organization/person that needs to decode it (Krippendorff, 1975). Due to these concerns, absent context affects knowledge transmission between the encoding and decoding stages (Alavi and Leidner, 2001), and as Cowan *et al.* (2000: 225) write: 'What is codified for one person or group may be tacit for another and an utterly impenetrable mystery for a third.' Knowledge embedded in one context is, therefore, of less value in another context, a devaluing caused by the problem of separating knowledge from one context and transferring it to another (Kogut and Zander, 1992; Grant, 1996; Hansen, 1999; Szulanski, 1996, 2000).

The network approach focuses on relationships among actors and the outcomes those relationships produce. This approach is central, since the subsidiary's relationships to counterparts in the local host environment are of particular importance for its daily business activities, e.g., production, sales activities and knowledge creation processes (Blanc and Sierra, 1999; McEvily and Zaheer, 1999; Andersson *et al.*, 2002; Frost *et al.*, 2002).

The reason for decontextualization depends on the degree of particularity in a relationship, as expressed by unique transfers in a dyadic relationship (Ford *et al.*, 1986). If a unique context is structured with the purpose of conducting business transactions, knowledge transfers become specifically and contextually embedded rather than standardized. The adaptation-process taking place between the subsidiary and its local partners further leads to context-specificity. Adaptations reflect a unilateral or mutual adjustment of attitudes, strategies, knowledge and knowledge transfer mechanisms in the network, manifested in modified products and processes (Forsgren *et al.*, 1995; Håkansson, 1982). However, such transfer skills take time to develop, and successful adaptation processes are often only achievable through a long-lasting relationship between counterparts. Empirical studies of MNC subsidiaries confirm that competence development relates to exchanges in business relationships and that 80 per cent of these relationships are with external counterparts of the MNC (Pahlberg, 1996; Andersson, 1997). The context-specific character of knowledge, and relationships creates obstacles for the transfer of knowledge and, at the same time, the knowledge may become less useful to other MNC units operating in a different context (Forsgren, 1997). Therefore, the more the subsidiary develops its knowledge in external relationships, the more of a transfer problem it is likely to encounter, due to the risk of context specificity. This argumentation leads to the following hypothesis (H1):

> **H1**: The higher the ratio of externally sourced knowledge, the less knowledge will be transferred to other MNC units.

A further pertinent issue is the extent to which the subsidiary spends its limited resources on internal knowledge transfer. Subsidiaries which invest these resources on internal knowledge transfer run the risk of not being able to allocate the resources or time that are necessary continuously to upgrade their own competences (Forsgren *et al.*, 2000).

In addition, small and medium-sized organizations typically do not have the capacity to either build close relationships with all 'partners' (Lane and Lubatkin, 1998), or to meet both internal and external institutional requirements (Kostova and Roth, 2002). The documentation of experiences, blueprinting of technology, etc., are time-consuming activities, seen as dead loss, since they do not directly lead to value creation in the subsidiary. Thus we propose:

> **H2**: The larger the size of the subsidiary, the more knowledge will be transferred to other MNC units.

In order to overcome the fragmentation problem, MNCs have intensified their efforts to make knowledge available across the organization. Strategies like 'total openness in internal communication', where everyone has full access to the organization's information and is aware of everyone else's repertoire (Grant, 1996), make a tremendous amount of information available to the individual. To exemplify this, Ernst and Young, a company which practises such a strategy, estimates that it has 1.2 million documents in its general unfiltered repository, 875,000 documents in its discussion databases, and 50,000 documents in comprehensive packs of material on specific topics (Wenger *et al.*, 2002). In practice, this situation is impossible to handle simply because of information overflow. As highlighted by Szulanski (2000), the first condition necessary to transfer a piece of knowledge is that the organization or person who needs the knowledge must be able to identify its location. The solution to the problem is to design an organizational structure that formally recognizes the existence and value of knowledge in subsidiaries (Bartlett and Ghoshal, 1989). By officially giving a mandate to the subsidiary in terms of dedicated resources for knowledge creation, headquarters creates an ability to enforce knowledge transfers among units, if needed (Birkinshaw, 1996; Galunic and Eisenhardt, 1996). Based on this discussion, we propose Hypothesis 3:

> **H3**: The higher the degree of formal recognition of subsidiary knowledge, the more knowledge will be transferred to other MNC units.

Some studies show a positive relationship between autonomy of subsidiaries and knowledge creation (Taggart, 1997; Taggart and Hood, 1999), whereas opposite results are found by Brockhoff and Schmaul (1996) and Ensign *et al.* (2000). The autonomous subsidiary is in a difficult position since it may experience a loss of bargaining power when transferring knowledge, whereas isolation might lead to knowledge hoarding (Szulanski, 1996; Husted and Michailova, 2002). Therefore, we propose:

> **H4**: The higher the degree of subsidiary autonomy, the less knowledge will be transferred to other MNC units.

Interchanges of products and resources with corporate entities are an inverse operation to autonomy (Garnier, 1982), and are important for internal embeddedness and integration in the MNC. Randøy and Li (1998) show a positive relationship between the flow of physical products and MNC integration. Both Randøy and Li (1998) and Gupta and Govindarajan (1994, 2000) advocate specific subsidiary roles to handle both inflows and outflows of products and resources. Thus we propose:

> **H5**: The larger the degree of interdependency, the more knowledge will be transferred from a subsidiary to other organizational units.

Data and method

A group of researchers established and launched the 'centre of excellence' project in May 1996[1] with the main purpose of investigating different subsidiary roles in MNCs. Researchers in the Nordic countries, United Kingdom, Germany, Austria, Italy, Portugal and Canada participated. One of the first issues was to obtain proper data for the project. In order to collect quantitative data on subsidiary roles, a questionnaire that could be applied in all the countries involved was constructed with the intention of identifying 'centres of excellence', explaining their existence and some of their effects. Approximately 80 per cent of the questionnaires were answered by subsidiary executive officers, while financial managers, marketing managers or controllers in the subsidiaries answered the remaining 20 per cent. The response rate varied between 20 and 55 per cent, depending on the country of investigation, and the quality of the data was quite high, with a general level of missing values of not more than 5 per cent. In total, answers were received from 2109 subsidiaries. Out of these, 25.1 per cent claimed that they conduct basic research. 54.3 per cent were engaged in developing products or processes and 67.1 per cent produced goods or services. As many as 95 per cent had marketing and sales activities, and about 86 per cent had activities in logistics, distribution or purchasing.

For all subsidiaries sampled, information exists on the subsidiary knowledge elements, notably subsidiary-level knowledge, knowledge sources, organizational context variables, and the extent to which knowledge has been transferred to other MNC units.

Part of the survey investigated the degree to which certain sources were influential on the subsidiaries' specific competences. The internal sources included: (1) MNC customers; (2) MNC suppliers; and (3) MNC R&D units. The external sources included: (A) customers (B) suppliers (C) distributors and (D) external R&D units. Some basic figures are presented in Table 10.1.

As shown in Table 10.1, external counterparts are very important sources for knowledge development in the subsidiaries. In fact, local customers have the highest score with a mean of 4.24. Competitors, suppliers and distributors have significant importance as external knowledge development sources as well. The conclusion is, therefore, that interactions with local counterparts constitute a significant source of knowledge development for most subsidiaries, implying that many subsidiaries are highly embedded in the local environment. This is in line with the study by Andersson *et al.* (2002), which concludes that subsidiaries mainly obtain their knowledge in external relationships through customers and suppliers.

Table 10.1 The importance of different sources for knowledge development in the subsidiary

	Mean	*Std. Dev.*
Internal source		
MNC customer	2.90	1.91
MNC supplier	2.87	1.83
MNC R&D unit	2.86	1.95
External source		
Customers	4.24	2.01
Suppliers	2.91	1.81
Distributors	2.65	1.83
External R&D unit	2.10	1.90

$N = 2109$. All figures based on a 1–7 Likert scale, where 1 = no importance and 7 = very significant importance. Internal sources have a standardized Cronbach Alpha Coefficient of 0.70. External sources have a standardized Cronbach Alpha Coefficient of 0.64.

Measures

All data was collected through the questionnaire and most variables are multi-item measures, which were measured using seven-point Likert scales. However, items such as the number of employees were measured using actual values. The following sections provide the exact wording used for questionnaire items.

Knowledge transfer

Our definition of knowledge transfer captures the application rather than the transfer *per se* of the subsidiary knowledge to other MNC units. Accordingly, the subsidiaries were asked to what extent the knowledge they control has been of use to other MNC units. Respondents indicated this on a seven-point Likert scale, where 1 was defined as 'to no use at all for other units' and 7 was defined as 'very useful for other units' for all of the following seven activities: research, development, production, marketing and sales, logistics, purchasing and HRM. Knowledge transfer is a multi-item construct calculated as the average score reported by respondents across these seven items.

Ratio of external knowledge

To calculate this item, a measure of 'external knowledge' (i.e. subsidiary-level knowledge built mainly from external knowledge inputs), which captures the importance of external parties, such as customers and suppliers, was first calculated. The inputs from external partners were measured by asking respondents to assess the impact of various external organizations on the development of the subsidiary's competencies, where 1 equalled 'no impact at all' and 7 equalled 'very decisive impact'. Four organizations were identified: external market customers, external market suppliers, specific distributors and specific external R&D units (see Table 10.1). Along the same vein, a measure of 'internal knowledge' (i.e. subsidiary-level knowledge built mainly from internal knowledge inputs) is calculated to capture knowledge developed through interaction with other MNC units. In order to measure this, respondents were asked to assess the impact of various internal organizations on the development of the subsidiary's competencies (on the same scale as above). Three organizations were identified: internal MNC customers, internal MNC suppliers and internal MNC R&D units (see Table 10.1). Finally, the ratio of external knowledge was calculated as the external knowledge measure divided by the internal knowledge measure.

Size of subsidiary

The size of the subsidiary is measured in two different ways: as the total sales of subsidiary (in thousands of US dollars), and as the level of investments in the subsidiary. The level of investments was measured by asking respondents to assess the level of investments in the subsidiary in the past three years, where 1 equalled 'very limited' and 7 equalled 'substantial'.

Recognition

Recognition was measured by asking respondents whether the subsidiary's competence was formally recognized by the MNC headquarters for each of the following seven activities: research, development, production, marketing and sales, logistics, purchasing and HRM. The subsidiaries were rated on a scale of 0 to 7 depending on the number of activities that were formally recognized by the MNC headquarters.

Autonomy

Based on the scale developed by Roth and Morrison (1992), respondents were asked to identify the level at which certain decisions were made, where 1 equalled foreign corporate (HQ), 2 equalled sub-corporate (e.g. division), and 3 equalled subsidiary level. The decisions were as follows: hiring top subsidiary management; entering new markets within the country; entering foreign markets; changes to subsidiary organization; introduction of new products/services; and approval of quarterly plans and schedules.

Interdependency

The level of intra-MNC trade is an indicator of the breadth and strength of the internal trade links. This measure was used as a proxy for interdependency, measured by two variables: the share of subsidiary sales going to other MNC units in 1996 (intra MNC export), and the share of subsidiary sales stemming from other MNC units, including both semi-products and final goods and services (intra-MNC import).

Controls

The *level of subsidiary knowledge*, which may have a significant impact on the level of knowledge transfer to other MNC units, was added as a control variable. The level of knowledge was measured by asking respondents to indicate the level of subsidiary knowledge on a 7-point Likert scale, where 1 equalled 'no knowledge' and 7 equalled 'substantial knowledge' for the seven activities: research, development, production, marketing and sales, logistics, purchasing and HRM.

Results and discussions

We have applied an OLS (ordinary least square) regression model to test all five hypotheses simultaneously. The results of the OLS-model are reported in Table 10.2. Numbers in parentheses represent *t*-values. The last column provides the variance inflation factor (VIF), which tests for multicollinearity in the data set. However, all VIF-values are far below the usual threshold of 6.0 for detecting multicollinearity, indicating that we have no problem of multicolleniarity in the data set.

As shown in Table 10.2, all of the hypothesized relationships are significant and with the expected signs. The context-specificity (as proxied by the ratio of external knowledge) and autonomy of the subsidiary have a negative impact, while size (measured by both the number of

Table 10.2 OLS regression: factors determining the level of knowledge transfer from the subsidiary to other MNC units (*t*-value in parentheses)

Model variable	*Knowledge transfer*	*Variance inflation*
Intercept	−0.40 (−2.73)***	0
Ratio of external knowledge	−0.09 (−2.83)***	1.07
Size	0.20	1.03
– Total sales	(4.70)***	
– Level of investment	0.25 (11.69)***	1.46
Recognition	0.17 (10.88)***	1.49
Autonomy	−0.10 (−1.76)*	1.07
Interdependency	0.04	1.12
– Intra-MNC import	(3.94)***	
– Intra-MNC trade (export)	0.12 (8.77)***	1.11
Control		
– Level of competence	0.40 (18.18)***	1.93
Adjusted R^2	**0.52**	
F-statistic	**275.96*** **	
N	**2109**	

*, ** and *** indicate 10%, 5% and 1% level of significance, respectively.

employees and the level of investment), recognition of knowledge and interdependency (both types of trade flows) have a significantly positive impact on the level of knowledge transfer.

These results confirm that obstacles and dilemmas are involved in the internal transfer of knowledge from subsidiaries to other MNC units. The more engaged the subsidiary is in external knowledge sourcing and the more decision rights (autonomy) it acquires, the less knowledge will typically be transferred to other MNC units. Therefore, there are some inherent dilemmas related to the subsidiary role as a bridgehead, since the subsidiary ought to acquire the knowledge externally, while too much emphasis on independence and external relationships may result in a lock-in on local context. However, the results also indicate that MNC headquarters can alter the incentive structure to hinder the subsidiary in becoming to locked onto the local context in their knowledge development. Some of the variables can, to some extent, be designed by the MNC headquarters, such as recognition of subsidiary knowledge, granted subsidiary autonomy, intensity of intra-MNC trade flows, and subsidiary resources. The MNC headquarters can apply these variables in order to facilitate internal knowledge transfer. In particular, recognition of subsidiary knowledge in the MNC seems to be important in this respect.

The inherent dilemma stemming from the context-specificity of subsidiary knowledge, and the lack of time and resources needed to make the knowledge transferable and usable in other MNC units is illustrated in the following quote:

> *When we have developed a new product, a new way of solving a specific problem, in principle we should transfer all the new knowledge to the other MNC units, but it is not possible. We can demonstrate the new solution in practice provides the other units with technical descriptions and all the knowledge we codify. We can transfer the knowledge on the final solution in a codified form, but we cannot transfer the knowledge we gained from all of the unsuccessful solutions to the problem. So, in a way, we are only transferring the solution to a specific problem and not the underlying competence. Therefore, as the underlying competence continues to reside in our subsidiary we are much better suited to modify and develop the product than other MNC units. (R&D manager in Danish subsidiary owned by a French MNC)*

As indicated in the quote, what is transferred to other MNC units is very often not the underlying competence, but rather applications of this competence in the form of a solution to one specific problem, perhaps codified in drawings, etc. It follows that the underlying competence remains in the subsidiary, so that the subsidiary will have the competence to make changes and adjustments to the suggested solution, while other MNC units will mainly gain the knowledge relevant to the specific solution. Applications, rather than competences themselves, are transferred because of both the fear of giving away the uniqueness and the context-specific nature of the competence, which makes transfer to other environments difficult. The loss of bargaining power and, in general, organizational position or mandates are strong incentives to protect uniqueness (Husted and Michailova, 2002).

In addition to unwillingness to participate in knowledge transfer, both competence development in the local context and involvement in corporate learning are time- and resource-consuming, which gives rise to a dilemma. On the one hand, the subsidiary may concentrate its time and resources slated for development on network relationships, usually external of the MNC. This 'autonomy' promotes high competence but is more related to the subsidiary's own activities without an extensive impact on other MNC units. On the other hand, the subsidiary can strive to integrate in the MNC, where it expands the internal relationships and is involved in corporate learning and strategic decisions. This is likely to give the subsidiary a significant power position with the possibility to influence MNC development. The danger of this 'integration' is that the underlying sources of knowledge development may be neglected, because scarce time and resources are spent on internal knowledge transfer rather than on knowledge creation.

The dilemma can be formulated as follows: How can the subsidiary engage in internal knowledge transfer in the MNC without giving away its uniqueness and neglecting its knowledge creation sources? The danger is that the subsidiary engages in extensive knowledge transfer without continuously developing its own knowledge base. In this case, the subsidiary will lose its uniqueness and its knowledge will decrease in value to other MNC units. Subsidiaries need to develop strategies continuously to nurture both the local business network and the corporate network, as it is dependent on the well being of both these networks.

Note

1. For more information on the 'centre of excellence' project, see Holm and Pedersen, 2000.

References

Alavi, M. and Leidner, D. (2001) Review: knowledge management and knowledge management systems: conceptual foundations and research issues. *MIS Quarterly*, 25: 107–36.

Andersson, U. (1997) *Subsidiary Network Embeddedness*. Doctoral Thesis no. 66. Uppsala: Department of Business Studies, Uppsala University.

Andersson, U., Forsgren, M. and Holm, U. (2002) The strategic impact of external networks: subsidiary performance and competence development in the multinational corporation. *Strategic Management Journal*, 23: 979–96.

Bartlett, C. A. and Ghoshal, S. (1989) *Managing Across Borders: The Transnational Solution*. Boston: Harvard Business School Press.

Birkinshaw, J. (1996) How multinational subsidiary mandates are gained and lost. *Journal of International Business Studies*, 27(3): 467–95.

Blanc, H. and Sierra, C. (1999) The internationalization of R&D by multinationals: a trade-off between external and internal proximity. *Cambridge Journal of Economics*, 23: 187–206.

Brockhoff, K. L. and Schmaul, B. (1996) Organization, autonomy, and success of internationally dispersed R&D facilities. *IEEE Transactions of Engineering Management*, 43: 33–40.

Cowan, R., David, P. A. and Foray, D. (2000) The explicit economics of knowledge codification and tacitness. *Industrial and Corporate Change*, 9: 211–53.

Ensign, P. C., Birkinshaw, J. M. and Frost, T. S. (2000) R&D centres of excellence in Canada. In U. Holm and T. Pederson (eds) *The Emergence and Impact of MNC Centres of Excellence: A Subsidiary Perspective*. Houndsmill, Basingstoke: Macmillan.

Ford, D., Håkansson, H. and Johanson, J. (1986) How do companies interact? *Industrial Marketing and Purchasing*, 1(1): 26–41.

Forsgren, M. (1997) The advantage paradox of the multinational corporation. In I. Björkman and M. Forsgren (eds) *The Nature of the International Firm: Nordic Contributions to International Business Research*. Copenhagen: Handelshøjskolens.

Forsgren, M., Holm, U. and Johanson, J. (1995) Division headquarters go abroad – a step in the internationalization of the multinational corporation. *Journal of Management Studies*, 32(4): 475–91.

Forsgren, M., Holm, U., Pedersen, T. and Sharma, D. (1999) The subsidiary role for MNC competence development: information bridgehead or competence distributor? Paper presented at the EIBA-conference, Manchester, December 9–13.

Forsgren, M., Johanson, J. and Sharma, D. (2000) Development of MNC Centres of Excellence. In U. Holm and T. Pedersen (eds) *Centres of Excellence*. London: Macmillan, pp. 45–67.

Frost, T. S., Birkinshaw, J. M. and Ensign, P. C. (2002) Centers of Excellence in multinational corporations. *Strategic Management Journal*, 23: 997–1018.

Galunic, C. D. and Eisenhardt, K. M. (1996) The evolution of intracorporate domains: divisional charter losses in high-technology, multidivisional corporations. *Organization Science*, 7(3): 255–82.

Garnier, G. H. (1982) Context and Decision-making autonomy in the foreign affiliates of US multinational corporations. *Academy of Management Journal*, 25: 893–908.

Grant, R. M. (1996) Toward a knowledge-based theory of the firm. *Strategic Management Journal*, 17(special issue): 109–122.

Gupta, A. K. and Govindarajan, V. (1991) Knowledge flows and the structure of control within multinational corporations. *Academy of Management Review*, 16: 768–92.

Gupta, A. K. and Govindarajan, V. (1994) Organizing for knowledge flows within MNCs. *International Business Review*, 3(4): 443–57.

Gupta, A. K. and Govindarajan, V. (2000) Knowledge flows within multinational corporations. *Strategic Management Journal*, 21(4): 473–96.

Hansen, M. T. (1999) The search-transfer problem: The role of weak ties in sharing knowledge across organization subunits. *Administrative Science Quarterly*, 44(1): 82–111.

Hedlund, G. (1986) The hypermodern MNC – a heterarchy? *Human Resource Management*, 25(12): 9–35.

Holm, U. and Pedersen, T. (eds) (2000) *Centres of Excellence*. London: Macmillan.

Husted, K. and Michailova, S. (2002) Diagnosing and fighting knowledge-sharing hostility. *Organizational Dynamics*, 31(1): 60–73.

Håkansson, H. (ed). (1982) *International Marketing and Purchasing of Industrial Goods*. Chichester: Wiley.

Kogut, B. and Zander, U. (1992) Knowledge of the firm, combinative capabilities, and the replication of technology. *Organization Science*, 3: 383–97.

Kostova, T. and Roth, K. (2002) Adoption of an organizational practice by subsidiaries of multinational corporations: institutional and relational effects. *Academy of Management Journal*, 45(1): 215–33.

Krippendorff, K. (1975) Some principles of information storage and retrieval in society. *General Systems*, 20: 15–35.

Lane, P. J. and Lubatkin, M. (1998) Relative absorptive capacity and interorganizational learning. *Strategic Management Journal*, 19: 461–77.

McEvily, B. and Zaheer, A. (1999) Bridging ties: a source of firm heterogeneity in competitive capabilities. *Strategic Management Journal*, 20: 1133–156.

Pahlberg, C. (1996) *Subsidiary–Headquarters Relationships in International Business Networks*. Doctoral thesis, Uppsala: Department of Business Studies, Uppsala University.

Randøy, T. and Li, J. (1998) Global resource flows and MNE network integration. In J. Birkinshaw and N. Hood (eds) *Multinational Corporate Evolution and Subsidiary Development*. Houndsmill, Basingstoke: Macmillan.

Roth, K. and Morrison, A. (1992) Implementing global strategy: characteristics of global subsidiary mandates. *Journal of International Business Studies*, 23: 715–35.

Szulanski, G. (1996) Exploring internal stickiness: impediments to the transfer of best practice within the firm. *Strategic Management Journal*, 17(winter): 27–43.

Szulanski, G. (2000) The process of knowledge transfer: a diachronic analysis of stickiness. *Organizational Behavior and Human Decision Processes*, 82(1): 9–27.

Taggart, J. H. (1997) R&D complexity in UK subsidiaries of manufacturing multinational corporations. *Technovation*, 17: 73–82.

Taggart, J. H. and Hood, N. (1999) Determinants of autonomy in multinational corporation subsidiaries. *European Management Journal*, 17: 226–36.

Wenger, E., McDermott, R. and Snyder, W. M. (2002) *Cultivating Communities of Practice: A Guide to Managing Knowledge*. Boston, MA.: Harvard Business School Press.

Part IV

Knowledge Governance and Business Development

11

Governing MNC Entry in Regional Knowledge Clusters

Mark Lorenzen and Volker Mahnke

Introduction

The governance of knowledge production and use is central to the theory of the MNC since its inception (Vernon, 1966; Hymer 1959; Buckley and Casson, 1976). One classical motive for FDI is to *exploit knowledge assets* developed at home. More recent research (Bartlett and Ghoshal, 1989; Hedlund, 1994; Cantwell, 1995; Florida, 1997; Grandstrand *et al.*, 1992; Kümmerle, 1999) stresses the need for cross-border learning. Thus, a central element in location decisions becomes *knowledge-exploration* in foreign markets.

Kümmerle (1999) and Patel and Vega (1999) observe that MNCs increasingly place R&D units in regional clusters to augment their knowledge bases. Frost (2001) suggests that such local knowledge sources are particularly important for explorative R&D that develops new knowledge for the MNC, rather than adapting existing knowledge developed elsewhere to local conditions. Several authors have illustrated this for a number of high-technology industries. For example, Almeida (1996) shows that the US subsidiaries of foreign MNCs draw heavily upon the technology of local companies. Shan and Song (1997) find that foreign MNCs make equity investments in US biotechnology firms with high levels of patent activity, thus sourcing firm-embodied technological advantages located away from the MNCs' host countries. Given the importance of international knowledge-seeking strategies, this paper is concerned with MNC entry in regional knowledge clusters.

Prior research (Caves and Mehra, 1986; Cho and Padmanabhan, 1995; Hennart and Park, 1993) examines the influence of a variety of factors (for instance, cultural distance, multinational experience, firm size, etc.) on *entry modes in product markets*. Results remain mixed and differ across

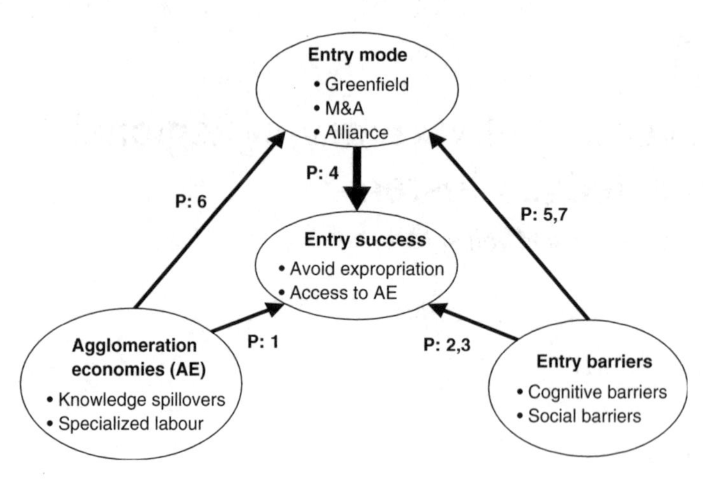

Figure 11.1 Knowledge cluster entry

contexts examined so that authors conclude that the state of the literature might be signified as under-developed (Chang, 1995; Melin, 1992). This is especially true with regards to MNC *entry modes in knowledge clusters*. As indicated in Figure 11.1, this paper develops propositions on (a) complications of entry strategies when the MNC seeks to explore knowledge in regional knowledge clusters, and (b) the determinants of entry mode choice (i.e. greenfield investments, strategic alliances, acquisitions) and different entry modes' impact on entry success in terms of protection against expropriation hazards and access to agglomeration economies.

We proceed as follows. First, we discuss several sources of agglomeration economies in knowledge clusters from which the entering MNC seeks to benefit. Second, we consider crucial factors influencing knowledge-seeking entry strategies. Finally, we discuss contingencies of entry mode choices in the light of sought-after agglomeration economies and expropriation hazards.

Sources of agglomeration economies in knowledge clusters

Within a range of industries, the bulk of innovative economic activity takes place within *regional clusters* (e.g. DeBresson, 1996; Staber, 1995; Steiner, 1998; Roelandt and Hertog, 1999, Porter, 1998; Schmitz, 1999). Examples include auto equipment in Detroit, US (Porter, 1998); textiles in Prato, North Italy (Piore and Sabel, 1984); surgical equipment in

Sialkot, Pakistan (Schmitz, 1999); and furniture in West Jutland, Denmark (Lorenzen and Foss, 2002). Regional clusters denote groups of interconnected firms (both suppliers, customers, and competitors) that operate in the context of local labour markets and private and public institutions (educational institutions; specialized public and private service suppliers; and labour market and employers' associations), specialized within a few and related economic activity areas. Some such clusters are distinctively *high-tech*. Here, technology leading firms, highly skilled labour (often engineers), and knowledge institutions (typically universities and research facilities) cluster together, often in urban areas. Good examples are microelectronics in Silicon Valley, US (Saxenian, 1994); equipment in Cambridge, UK (Keeble and Wilkinson, 1999); and ICT in Bangalore, India (Patibandla, 1998). Such clusters will be referred to as *knowledge clusters*. Regional concentration of economic activity in particular knowledge clusters yields *agglomeration economies* – essentially positive externalities that benefit locally present firms in three forms.

Vertical knowledge spillovers

First, some of these agglomeration economies are due to direct ties between vertically related firms and result in *external scale and scope economies*. Firms in local knowledge clusters benefit from cost-efficiency and flexible specialization due to their access to co-specialized local suppliers that enable a firm to produce in shifting volumes while offering broad and flexible product ranges (see e.g. Piore and Sabel, 1984; Scott and Storper, 1986). Such vertical relations between firms in a cluster often lead to *vertical knowledge spillovers*, as demands and feedback from customers and specialized suppliers may push and pull incremental upgrading of a firm's knowledge base and lead to process or product innovations (von Hippel, 1988; Lundvall, 1988). This process is eased by the geographical proximity of cluster firms, because some types of knowledge – spanning from subjective advice to technical knowledge – are best transferred through direct observations and face-to-face interactions. If proximity allows for frequent face-to-face meetings and on-site observations in vertical relations, knowledge exchange costs are reduced and learning from others is eased.

Horizontal knowledge spillovers

Second, agglomeration benefits are due to *horizontal knowledge spillovers*. For example, clustered firms monitor and absorb technological advance generated by other horizontally related cluster incumbents (Lorenzen and Foss, 2002). Such knowledge spillovers are not restricted to direct business relations such as strategic alliances between firms but importantly result from common 'third party' relations. For example, common cognitive platforms emerge (Lorenz and Lazaric, 1998) due to social interaction of employees or managers (e.g. in both professional and social life) across cluster firms, to ease mutual monitoring and learning. Information stemming from monitoring of competitors' experiments with markets and technologies is one important input to a firm's own knowledge production and can result from both planned environment scanning and accidental observations (e.g. gossip).

Monitoring is facilitated in clusters, because indirect relations between people are more frequent with geographical proximity. Whereas direct relations between firms (at least, in the successful cases) allow for in-depth transfer of knowledge, indirect relations allow firms to monitor a wide range of information (Granovetter, 1973, 1982). This information also includes what is not expected nor searched for, which may have a greater potential for inspiring change and innovation in firms than 'the provincial news and views of their close friends' (Granovetter, 1982: 106). In Marshall's (1891) words, information may be simply 'in the air' within a cluster.

Specialized labour and institutions

A third type of agglomeration economies arises due to a *specialized local labour pool* and labour market institutions; specialized local public or semi-public R&D facilities (such as industrial development boards or think tanks); educational organizations (such as universities); and a range of other public and private services (e.g. finance). For example, local concentration of economic activity within a few related areas is often accompanied by specialization of labour. A high demand for particular qualifications helps in targeting public (and sometimes private) educational institutions and encourages political will to expand upon particular educational activities through, for example, investments in technical schools and universities. In addition, universities and other research institutions facilitate technology transfer by acting as partners in technologically development projects. In-job training further augments the skill level in the local labour force. Finally, good employment opportunities mean that skilled labour tends to migrate to the cluster (at least, to many urban clusters).

As a result, a specialized local labour market accommodates the cluster's dominant type of economic activity. Such a labour market caters to the need of cluster firms and acts as a mechanism for transferring knowledge between firms. Educational institutions augment the pool of qualified labour in a cluster because they provide a platform where business managers or employees of different firms in a cluster continuously meet, exchange information and build social ties – at times originating from joint days in a local school, college, or university. Such social ties also often facilitate trust between people. In addition, they create a common codebook for communication as shared social conventions and frames of reference enable normative and cognitive integration that eases interaction and communication between people and firms (Scott, 1995; Schmitz, 1999).

In sum, firms located in a knowledge cluster can potentially benefit from external scale and scope economies through engaging in vertical linkages, from identifying, accessing, and exploiting knowledge in horizontal linkages, as well as from social capital present in direct and indirect network relations and a specialized labour market. Thus, a knowledge-seeking MNC may seek to benefit from such agglomeration economies available in local knowledge clusters. Thus, our first proposition:

> **P1**: A knowledge-seeking MNC will benefit more from cluster entry the greater the access to (a) specialized labour, (b) vertical knowledge spillovers, and (c) horizontal knowledge spillovers.

Complications of entry in knowledge clusters

What are crucial factors influencing a MNC's knowledge-seeking entry strategy? Whether the MNC benefits from agglomeration economies present in a knowledge cluster depends, *inter alia*, on two central factors: (a) cognitive entry barriers, and (b) social entry barriers.

Cognitive entry barriers

A MNC might fail to benefit from agglomeration economies in knowledge clusters if it lacks absorptive capacity (Cohen and Levinthal, 1990). For example, Frost (2001) suggests that a technologically strong MNC will benefit more from entering into a knowledge cluster. Technological strength allows an MNC to absorb technological knowledge and innovative ideas that weaker firms in the cluster might develop but are unable to exploit by themselves. Cantwell and Janne (1999) agree, when they argue that MNCs with a home base in 'centres of excellence' have better possibilities for tapping knowledge in clusters around the world.

While a strong technological absorptive capacity increases the possibility of identifying and understanding ideas of others, incumbent firms in a knowledge cluster can also imitate technological advance of technologically strong MNCs. Shaver and Flyer (2001) propose that a technologically strong MNC may sometimes have disincentives to enter into clusters, because it may benefit less from knowledge spillovers compared to weaker incumbent firms. A knowledge-seeking MNC may, in fact, suffer diseconomies of agglomeration, if incumbent competitors imitate explicit technological know-how faster than entrants in knowledge clusters can pick up and use ideas from incumbents.

Even if a knowledge-seeking MNC is able to identify valuable knowledge in a cluster, to put this knowledge to use is far from easy. Because they are embedded in network relations cluster incumbents may have a better understanding of technological knowledge that originates within the cluster (for example, they may have participated to inventing technologies). An entering MNC may experience some 'liabilities of foreignness', an effect which only decreases slowly after entry. This is because technological advance might be based on socially complex tacit knowledge or may be 'coded' in ways that are comprehensible to only those firms that share 'codebooks', i.e. particular ways of formulating it, contextualizing it, sharing it and using it (Lorenzen, 1998; Cowan *et al.*, 2000; Lissoni, 2001). Thus, it may take a substantial period of time for a newcomer firm to build a codebook that allows it to identify and understand local knowledge and put it to productive use. Possessing a high technological absorptive capacity may be helpful in this respect, but local knowledge production and ways of sharing it can be highly endemic and take more than technical skills to master. Consequently, knowledge appropriation problems are particularly pronounced for technological weak entrants that lack both technological and cluster-specific absorptive capacities. Based on our discussion above we suggest the following refutable proposition:

> **P2**: A knowledge-seeking MNC will benefit more from cluster entry (a) the greater its technological absorptive capacity, (b) the lower the imitability of the MNC's knowledge assets by competitors, and (c) the faster it can develop cluster-specific absorptive capacity by becoming embedded in knowledge-producing cluster relations.

Social entry barriers

When a MNC enters a cluster, agglomeration economies – and their distribution – often shift. Social entry barriers constitute a cost to an entering MNC. If incumbents meet MNC entry with suspicion, it may be difficult for it to become part of a network of indirect relations, in particular if local cluster networks are 'identity based' (Hite and

Hesterly, 2001). In such networks, social conventions and ambiguous ways of qualifying for trust and acceptance in cluster relations complicates MNC entry. A MNC, in addition, may meet resistance by being excluded from indirect relations, such as membership to industrial associations or social clubs, and incumbent firms may 'hide' social conventions or principles for communication, possibly allowing the newcomer into networks, but refraining from initiating the entrants on how, where and when local information sharing takes place. Hence, social entry barriers may constitute a serious barrier to reaping agglomeration economies that the knowledge seeking MNC seeks to obtain.

Social entry barriers depend on whether incumbents see the entering MNC as a threat or opportunity, the structure of the cluster, as well as labour market competition.

For example, a technologically strong MNC may bring complementary yet related knowledge to the cluster to increase the cluster's knowledge diversity. If such knowledge spills over to incumbents, they may benefit from a higher potential for innovation because greater diversity of knowledge leads to new knowledge combination possibilities. On the other hand, if the relative absorptive capacity of incumbents is low they may have more to loose than to gain from MNC entry. Accordingly, social entry barriers that the MNC faces depend on whether incumbents see the entering MNC as a threat or opportunity.

If there is competition for talent in a knowledge cluster, MNC entry is likely to increase labour market rivalry and accordingly meet resistance by incumbents. For example, if a cluster experiences more entry than local educational institutions and immigration of qualified labour can accommodate, competition for qualified labour may be intense between local firms. Latecomers may experience difficulties of attracting qualified labour but, as Patibandla (1998) points out, well-reputed MNC entry can drain labour supply for cluster incumbents through attracting key employees from incumbent firms.

Moreover, if there are dominant incumbent firms that strategically intervene, invest and centrally coordinate inter-firm relations in the cluster – what Rugman and D'Cruz (2000) coin *flagship firms* – these will be less interested in MNC entry, if this challenges their dominating power positions. On the other hand, MNC entry in a more 'symmetrical' cluster is likely to meet less resistance.

Based on the foregoing discussion we suggest the following refutable proposition:

P3: A knowledge-seeking MNC will benefit more from cluster entry (a) the lower local labour market rivalry, (b) the more symmetric the knowledge cluster configuration prior to entry, (c) the greater the perceived contribution of the entrant by incumbent firms, and (d) the less developed the cluster's social identity.

Entry mode choice and its implications

How do alternative entry modes influence the knowledge-seeking MNC's ability to benefit from entry into knowledge cluster? Our discussion above suggests that the success of different entry modes depends, among other factors, upon their relative advantages and dis-advantages for tapping into various kinds of agglomeration benefits (e.g. vertical and horizontal networks and the local pool of specialized labour), and hence, upon how entry modes are able to lower cognitive and social entry barriers. In addition, we shall claim that an important additional factor is how different entry modes facilitate necessary protection against knowledge expropriation risks. Thus:

> **P4**: A knowledge-seeking MNC's entry mode influences entry success in terms of protection against knowledge expropriation hazards and access to agglomeration economies.

Entry mode choice and expropriation risk

Knowledge-seeking FDI in knowledge cluster seeks to explore competitively valuable knowledge for future use (March, 1991; Kümmerle, 1999). An entering MNC seeks to protect competitively novel discoveries from imitation by cluster incumbents. The need to protect knowledge might be particularly strong where the entering MNC faces a situation where local incumbents' technological strength allows them to imitate the MNC's knowledge base quickly while the MNC's ability to absorb local knowledge remains limited. To the extent that explorative knowledge is process-based rather than product-based and at an early stage of the research pipeline, legal protection though patents is often prevented (e.g. Mansfield, 1984; Barkema and Vermeulen, 1998). Protections of valuable process knowledge through hierarchical governance forms can substitutes for a lack of legal protection (Liebeskind, 1997).

Strategic alliances remain less pronounced in the context of knowledge cluster entry. There is, however, a vast amount of literature dealing with learning alliances in general (Lam, 1997; Hamel, 1991; Lyles and Stalk, 1996; Inkpen and Beamish, 1997). While strategic alliance may provide an opportunity to learn, there are also substantial knowledge expropriation risks involved. Hamel (1991) argues that learning races between partner firms often pose a substantial risk of knowledge leakage. Inkpen and Beamish (1997) suggest that there may be incentive conflicts between partner firms. This may be especially the case if R&D is mainly established to develop process rather than product innovation, because the latter can be better protected by law compared to the former (Williamson, 1996). If a MNC judges the risk of imitation by potential local partner firms as substantial, it may refrain from strategic alliances.

More protection against knowledge expropriation may be afforded by knowledge-seeking entry by greenfield investment or acquisition. Several studies suggest that knowledge-seeking MNCs choose greenfield investment (e.g. setting up a R&D subsidiary in the cluster from scratch) rather than acquisition as entry mode (e.g. Mansfield, 1984; Kümmerle, 1999). Others suggest that MNCs use acquisition when they are technologically weak (Granstrand and Sjolander, 1990) or face a deteriorating knowledge base (Vermeulen and Barkema, 2001), while others observe an increasing trend of knowledge-seeking FDI by acquisitions in R&D (Andersson, 1997).

While the literature agrees that more hierarchical governance forms afford higher protection against expropriation risks, the distinction between knowledge-seeking entry by acquisition and greenfield investment is less clear. Kümmerle (1998) indicates that knowledge seeking entry by acquisition is riddled with evaluation problems arising when knowledge cannot be patented or is embodied in humans. Moreover, Hennart and Park (1993) suggest that firms with strong technological capabilities benefit less from acquisition in terms of acquired firm-specific assets and tend to choose greenfield investments instead. Finally, greenfield investments provide similar protection of intellectual assets while avoiding the costs of post-merger integration (Jemison and Sitkin, 1986; Datta, 1991), in particular, when the ratio of undesired to desired assets is high (Hennart and Park, 1993). In sum, then, we suggest the following refutable proposition:

> **P5**: A knowledge-seeking MNC will benefit more from cluster entry through hierarchical entry modes the higher expropriation risks are. The greater its technological strength, a knowledge-seeking MNC will benefit more from cluster entry through greenfield than through acquisition.

Entry modes and agglomeration economies

Entry mode decisions by a knowledge-seeking MNC are not only determined by concerns about knowledge expropriation. In addition, the relative advantages and disadvantages of entry modes (e.g. greenfield, strategic alliance, acquisition) for tapping into various kinds of agglomeration benefits (e.g. vertical, horizontal knowledge spillovers and, specialized labour) will influence entry decisions.

The studies mentioned above do not comprehensively control for social entry barriers and only partly account for cognitive entry barriers through focusing on the general technological strength of the entering MNC (but not for cluster-specific absorptive capacity). However, especially when social entry barriers are high, to shorten the process of accessing locally embedded collective knowledge assets available in a cluster it may be an advantage for an entering firm to 'plug into' the knowledge cluster via acquiring an incumbent. For example, an acquisition can provide instant access to network relations and specialized labour, while a greenfield investment needs to gradually develop its own network relations and go through time extensive hiring processes. In addition, if labour market constraints obtain, acquisitions can despite eventual complications of post-merger integration benefit the knowledge-seeking MNC because an acquisition provides instant access to specialized labour. To be sure, in this case human resources must be effectively bound to the entering MNC (Almeida and Kogut, 1999).

If strong incumbents pre-empt vertical and horizontal network relations (Dyer and Nobeoka, 2000), as is often the case in clusters governed by a flagship firm, there is little choice for the knowledge seeking MNC but to acquire an incumbent in the knowledge cluster or else use an alliance partner's established relations to overcome social entry barriers. By doing so, the entering MNC will increase its chances to gain access not only to skilled employees (including their social network), but also to the acquired firm's pre-established opportunities to benefit from both vertical and horizontal knowledge spillover. By implication we suggest the following refutable proposition:

> **P6**: When social entry barriers are high, knowledge-seeking MNC entry into knowledge clusters through acquisition will be more beneficial in terms of access to agglomeration benefits than entry through strategic alliances and greenfield investment.

If social entry barriers are low, for example, because there are few labour market constraints and the power distribution in the cluster is symmetric, entry is eased and expected entry cost imposed by incumbents will be lower. This is, essentially, because incumbents have little reason for retaliation or hostile behaviour to crowd out the entering MNC from direct and indirect network relations (be they vertical or horizontal). In addition, if clusters are symmetric and power is dispersed, resistance to entrants is harder to coordinate among incumbents in the knowledge cluster. If expropriation hazards are also low and the entering MNC's knowledge base requires less protection through hierarchical modes of entry, strategic alliances may act as a platform for learning local code books, developing cluster specific absorptive capacity, and gaining access to agglomeration economies. By implication we suggest the following refutable proposition:

> **P7**: When social entry barriers and expropriation hazards are low, knowledge-seeking MNC entry in knowledge clusters through strategic alliances increases the speed of developing local code-books and cluster specific absorptive capacity.

Conclusions

Compared to the vast literature on foreign entry in output markets, the literature on MNC entry in knowledge clusters is less well developed. This paper has outlined several propositions that inform research and managerial decision-making on the determinants of knowledge-seeking entry by a MNC in knowledge clusters. As is evident from the forgoing discussion, a knowledge-seeking MNC must make multiple considerations when seeking to tap into agglomeration economies present in knowledge cluster. In a nutshell, a MNC chooses entry modes that (a) facilitate necessary protection against knowledge expropriation risks while (b) maximizes benefits from access to agglomeration economies in the cluster.

We have separately examined and developed propositions for each aspect of entry modes, drawing attention to social and cognitive entry barriers that complicate cluster entry. As our propositions indicate, the threat of unwanted knowledge leakage should lead to more hierarchical forms of entry (e.g. greenfield and acquisitions), whereby entry by greenfield investment appears to be generally preferred if is central protection of intellectual assets at lower costs.

The discussion on agglomeration economies in knowledge clusters suggests that a MNC should also consider how entry mode choices influences its ability to access of various types of agglomeration economies and how fast this access can be achieved given social and cognitive entry barriers. Here we have argued that acquisitions might be the preferred entry mode that the knowledge-seeking MNC employs if it aims at speedily accessing agglomeration benefits. This is particular relevant if social entry barriers are high. As far as the problem of cognitive entry barriers is concerned, we suggest that even a technologically strong MNC with high general absorptive capacities may lack cluster specific absorptive capacity, i.e. relevant code books for appropriating localized knowledge, and this may be another reason for acquisition.

The essay has added to the MNC literature by introducing agglomeration economies as a factor explaining the choice of entry modes by MNCs that enter into knowledge clusters. We seek to stimulate empirical research on a knowledge-seeking MNC's governance choice through pointing to the potential advantages and disadvantages of different entry modes that should be taken into consideration in the investigation of a MNC's knowledge sourcing strategy.

By placing our research question in the context of local knowledge clusters, we identified factors (social and cognitive entry barriers) that have not been considered in the debate so far, but seem to be of growing empirical importance in the thrust of the MNC to tap local knowledge sources to advance innovativeness and foster technological edge vis-à-vis competitors.

References

Almeida, P. (1996) Knowledge sourcing by foreign multinationals: patent citation analysis in the US semiconductor industry. *Strategic Management Journal*, 17: 155–65.

Almeida, P. and Kogut, B. (1999) The localization of knowledge and the mobility of engineers in regional networks. *Management Science*, 45(7): 905–18.

Andersson, T. (1997) Internationalization of research and development, *Economics of Innovation and New Technology*, 3: 77–98.

Barkema, H. and Vermeulen, F. (1998) International expansion through start up or acquisition: a learning perspective. *Academy of Management Journal*, 41: 7–26.

Bartlett, C. A. and Ghoshal, S. (1989) *Managing Across Borders: The Transnational Solution*. Boston, MA: Harvard Business School Press.

Buckley, P. and Casson, M. (1976) *The Future of Multinational Enterprises*. London: Holmes & Meier.

Cantwell, J. A. (1995) The globalisation of technology: what remains of the product cycle model? *Cambridge Journal of Economics*, 19: 155–74.

Cantwell, J. and Janne, O. (1999) Technological globalization and innovative centres: the role of corporate technological leadership and locational hierarchy. *Research Policy*, 28: 119–44.

Caves, R. E. (1971) International corporations: the industrial economics of foreign investment. *Economica*, 38: 1–27.
Caves, R. E. and Mehra, S. (1986) Entry of foreign multinationals into US manufacturing industries. In M. Porter (ed.) *Competition in Global Industries*. Boston, MA: Harvard Business School Press, 459–81.
Chang, S. J. (1995) International expansion strategy of Japanese firms: capability building through sequential entry. *Academy of Management Journal*, 39: 383–407.
Cho, K. R. and Padmanabhan, P. (1995) Acquisition versus new venture: the choice of foreign stablishment mode by Japanese firms. *Journal of International Management*, 1(3): 255–85.
Cohen, W. M. and Levinthal, D. (1990) Absorptive capacity: a new perspective on learning and innovation. *Administrative Science Quarterly*, 35: 128–52.
Cowan, R., David, P. and Foray, D. (2000) The explicit economics of knowledge codification and tacitness. *Industrial and Corporate Change*, 9(2): 211–54.
Datta, D. (1991) Organization fit and acquisition performance: effects of post-acquisition integration. *Strategic Management Journal*, 12(4): 281–98.
DeBresson, C. (1996) *Economic Interdependence and Innovative Activity*. Cheltenham, UK: Edward Elgar.
Dyer, J. and Nobeoka, N. (2000) Creating and managing a high-performance knowledge-sharing network: the Toyota case. *Strategic Management Journal*, 21(3): 345–65.
Florida, R. (1997) The globalization of R&D: results of a survey of foreign-affiliated R&D laboratories in the USA. *Research Policy*, 26: 85–103.
Frost, T. S. (2001) The geographic sources of foreign subsidiaries' innovation. *Strategic Management Journal*, 22(2): 101–23.
Granstrand, O. and Sjolander, S. (1990) Strategic technology issues in Japanese manufacturing industry. *Technology Analysis and Strategic Management*, 1(3): 259–72.
Granstrand, O., Bohlin, E., Oskarsson, C. and Sjoberg, N. (1992) External technology acquisition in large multi-technology corporations. *R&D Management*, 22(2): 111–33.
Granovetter, M. (1973) The strength of weak ties. *Americal Journal of Sociology*, 78(6): 1360–80.
Granovetter, M. (1982) The strength of weak ties: a network theory revisited, in *Social Structure and Network Analysis*, Marsden and Lin (eds). Beverly Hills: Sage.
Hamel, G. (1991) Competition for competence and interpartner learning within international strategic alliances. *Strategic Management Journal*, 12: 83–103.
Hedlund, G. (1994) A model of knowledge management and the n-form corporation. *Strategic Management Journal*, 15: 73–90.
Hennart, J. F. and Park, Y. R. (1993) Greenfield *vs* Acquisition. *Management Science*, 39: 1054–70.
Hippel, E. von (1988) *The Sources of Innovation*. New York: Oxford University Press.
Hite, J. and Hesterly, W. (2001) The evolution of firm networks: from emergence to early growth of the firm. *Strategic Management Journal*, 22(3): 275–86.
Hymer, S. (1959) *The International Operations of National Firms: A Study of Direct Investment*. PhD dissertation, MIT.
Inkpen, A. C. and Beamish, P. W. (1997) Knowledge, bargaining power and international joint venture instability. *Academy of Management Review*, 22(1): 177–202.

Jemison, D. B. and Sitkin, S. B. (1986) Corporate acquisitions: a process perspective. *Academy of Management Review*, 11: 145–63.

Keeble, D. and Wilkinson, F. (1999) Collective learning and knowledge development in the evolution of regional clusters of high technology SMEs in Europe. *Regional Studies*, 33(4): 295–304.

Kogut, B. and Singh, H. (1988) The effect of national culture on the choice of entry mode. *Journal of International Business Studies*, Fall: 411–32.

Kümmerle, W. (1999) Foreign direct investment in industrial research in the pharmaceutical and electronics industries – results from a survey of multinational firms. *Research Policy*, 28: 179–93.

Lam, A. (1997) Embedded firms, embedded knowledge: problems of collaboration and knowledge transfer in global cooperative ventures. *Organisation Studies*, 18(6): 344–70.

Liebeskind, J. (1997) Keeping organizational secret: protective institutional mechanisms and their costs. *Industrial and Corporate Change*, 6(3): 623–63.

Lissoni, F. (2001) Knowledge codification and the geography of innovation: the case of Brescia mechanical cluster. *Research Policy*, 30(9): 1479–1500.

Lorenz, E. H. and Lazaric, N. (eds) (1998) *Trust and Economic Learning*. Cheltenham: Edward Elgar.

Lorenzen, M. (1998) Communicating and learning a basis for trust: Danish furniture producers in the Salling district, in Lorenzen (ed.) *Specialisation, Low-tech Competitiveness and Localised Learning: Six Studies on the European Furniture Industry*. Copenhagen: CBS Press.

Lorenzen, M. and Foss, N. (2002) Cognitive coordination, institutions, and clusters: an exploratory discussion. In Th. Brenner (ed.) *The Influence of Co-operations, Networks and Institutions on Regional Innovation Systems*. Cheltenham: Edward Elgar.

Lyles, M. A. and Salk, J. E. (1996) Knowledge acquisition from foreign parents in international joint ventures. *Journal of International Business Studies*, 27(5): 877–904.

Lundvall, B.Å. (1988) Innovation as an interactive process: from user–producer interaction to the national system of innovation. In G Dosi *et al.* (eds) *Technical Change and Economic Theory*. London: Pinter.

Mansfield, E. (1984) R&D and innovation: some empirical findings. In Z. Gilches (ed.) *R&D, Patents, and Productivity*. Chicago: University of Chicago Press.

March, J. (1991) Exploration and exploitation in organizational learning. *Organization Science*, 2 (Special issue): 71–87.

Marshall, A. (1891) *Principles of Economics*. London: Macmillan.

Melin, L. (1992) Internationalization as a strategy process. *Strategic Management Journal*, 13: 99–118.

Patel, P. and Vega, M. (1999) Patterns of internationalization of corporate technology: Location *vs* home country advantages. *Research Policy*, 28: 145–55.

Patibandla, M. (1998) Structure, organizational behavior and technical efficiency: the case of an Indian industry. *Journal of Economic Behavior and Organization*, March, 34(3): 419–34.

Piore, M. J. and Sabel, C. F. (1984) *The Second Industrial Divide: Possibilities for Prosperity*. New York: Basic Books.

Porter, M. (1998) Clusters and the new economics of competition. *Harvard Business Review*, 76(6): 77–90.

Roelandt, T. and Hertog, P. (eds) (1999) *Cluster Analysis and Cluster-based Policy: New Perspectives and Rationale in Innovation Policy*. Paris: OECD.

Rugman, A. and D'Cruz, J. (2000) *Multinationals as Flagship Firms*. Oxford: Oxford University Press, 2000.

Saxenian, A. (1994) *Regional Advantage*. Cambridge, MA: Harvard University Press.

Schmitz, H. (1999) Collective efficiency and increasing returns. *Cambridge Journal of Economics*, 23: 465–83.

Scott, R. (1995) *Institutions and Organizations*. Thousand Oaks, CA: Sage.

Scott, A. J. and Storper, M. (eds) (1986) *Production, Work and Territory: The Geographical Anatomy of Industrial Capitalism*. London: Allen & Unwin.

Shan, W. and Song, J. (1997) Foreign direct investment and the sourcing of technology advantage: an evidence from the biotechnology industry. *Journal of International Business Studies*, 28: 267–84.

Shaver, J. M. and Flyer, F. (2001) Agglomeration economies, firm heterogeneity, and foreign direct investment in the United States. *Strategic Management Journal*, 21(12): 1175–93.

Staber, U. (eds) (1995) *Business Networks: Prospects for Regional Development*, Berlin: De Gruyter.

Steiner, M. (ed.) (1998) *Clusters and Regional Specialisation: On Geography, Technology, and Networks*. London: Pion.

Vemeulen, F. and Barkema, H. (2001) Learning through acquisition. *Academy of Management Journal*, 44(3): 457–76.

Vernon, R. (1966) international investment and international trade in the product cycle. *Quarterly Journal of Economics*, 80: 190–207.

Williamson, O. (1996) *The Mechanisms of Governance*. New York: Free Press.

12

Learning and Networking in Foreign-Market Entry of Service Firms

Anders Blomstermo and D. Deo Sharma

Data system goes abroad

In 1968, the managing director of a British consulting company in Scandinavia started to develop an idea of how to increase sales with the help of IBM. The idea was developed from experience in selling management consulting services. However, cooperation with a hardware supplier opposed the policy of his company, which wanted to be as independent as possible. But the managing director stood by his view. In 1976, Data system was formed and Data system initiated a long-term cooperation with IBM, which has been reinforced in the meantime. Data system offers software, hardware and services to industrial customers, and is one of the world's largest suppliers of IBM AS/400-based software and services. Of Data system's total turnover, services make up about 75 per cent.

In 1995, the company had agents in 27 countries, subsidiaries in 16 countries, and 30 years of international experience. Data system has several large international clients. Foreign establishment started in 1981 in the neighbouring Nordic countries. IBM in the Nordic countries was consulted to find appropriate candidates for purchase. Through IBM, Data system received important information and a list of potential candidates and agents. Data system actively sought international opportunities.

In 1996, a large number of international customer relationships had been established. Portugal and Spain were given high priority. In Portugal, Data system had performed two installations via Data system's Belgian subsidiary. Data system's German agent was personally acquainted with the managing director of a consultant company in Portugal. He had established the contacts while working for his previous employer IBM. The first business contact was made in 1996, when the manager of the Portuguese company was in Stockholm studying Data system's products. After that, consultants from Portugal came to Stockholm for training.

The case of Portugal is a prototype with respect to the establishment of an agent. Establishment was made quickly (10 months) and it has not been expensive for Data system to start the business. Cultural and language problems have been small. This is due, among other things, to the agent having had prior experience of project handling, though no previous knowledge in the Data system area of application. The Portuguese agent was interested in learning about Data system's products and in making local product adaptations.

In Spain 1997, Data system tried to find a suitable agent via IBM, but found nothing of interest. Data system's knowledge of Spain was perceived to be high as three installations had previously been performed there for international customers. The success in Portugal also shaped and raised Data system's expectations for Spain. Data system received a personal recommendation from someone, with extensive experience of the agent, who had earlier worked at Volvo in Spain, and a contract was signed. Two years later this agent went bankrupt. The cultural differences had been large, resulting in misunderstanding between Data system and the agent. The language was also a problem, since few people in Data system speak Spanish and the Spaniards did not speak English well. The agent showed little interest in selling Data system's products. No installations were made.

Establishment in the Far East started in 1998, when a large international client wished to have installations in Asia. Four countries, Hong Kong, Indonesia, Singapore, Malaysia, are now served by the same company.

The case highlights internationalization, experiential knowledge and relationships. Data system's first foreign-market entries occurred in Scandinavia, followed by gradual establishment in more distant markets. At the start of the internationalization, Data system's foreign market knowledge and experience were limited. Data system drew on the experience of a British consulting company, international customers and IBM when collecting and interpreting information on foreign markets and suggestions on buy-out candidates and cooperating partners. By a process of trial and error, Data system gained knowledge and confidence in finding solutions. This has been reinforced by cooperation with local customers and with the hardware supplier, IBM. Through these business relationships, Data system has acquired foreign experience. In the beginning of the internationalization, Data system actively sought business relationships in order to go abroad. Then, international customers placed demands on Data system to make installations in several countries. Data system gained international experience and uncertainty decreased. Knowledge and experience of international markets and clients has accumulated through learning by doing.

A number of studies during the last decades have supported ideas and concepts from the behavioural models of internationalization. Knowledge accumulation and experiential learning has become a critical concept in internationalization research (Axelsson and Johanson, 1992; Chang, 1995; Eriksson, Johanson, Majkgård and Sharma, 1997; Erramilli, 1991). Studies show that performance is closely related to experiential learning (Barkema, Bell and Penning, 1996). The case also shows the need for internationalization models that can capture the early phase of internationalization. In addition, the old models of the internationalization process of the firm have also been criticized (Benito and Gripsrud, 1992; Petersen and Pedersen, 1997).

The Data system case is an example of the phenomenon studied in this chapter – knowledge and learning and networking in the internationalization process of service firms. The purpose of the chapter is to investigate experiential knowledge learning and knowledge management in the selection of foreign market strategies by (service) firms. What are the reasons for firms going abroad? Which foreign market strategy do firms pursue? What influences the selection of foreign markets? Finally, what are the consequences of the above on knowledge management in service firms? A distinction between market-seeking and client-following strategies is proposed. The empirical data for the papers is collected from service industries firms.

The chapter starts with a review of the literature on foreign market strategies and foreign market involvement, after which market-seeking and client-following strategies are described. In the literature review we combine the internationalization process model and the network theory. Then hard and soft services are discussed. Thereafter, the data are presented. Then we discuss the existing literature on the process approach and networks vis-à-vis foreign market strategies, and present data on how service firms actually enter foreign markets. Finally, implications for knowledge management in (service) firms and future research are discussed.

Internationalization strategies

The role of international firms is increasing in the context of international production of goods and services. Firms operating in areas such as banking, finance, insurance, health care and professional services, are increasingly able to exploit business opportunities abroad. This trend toward international expansion is expected to accelerate in the years to come. Thus, there is a real need for more studies on the entry of firms into foreign markets.

Studies have proposed a contingency entry mode and suggest that the selected foreign market should conform to country, industry, and firm-specific factors (Stopford and Wells, 1972; Zejan, 1990). Together the researchers found that location, ownership and firm-specific advantage affect foreign market selection. These studies have, however, supplied a cross-sectional picture of foreign market selection. The current literature distinguishes between the Uppsala (U) model and the innovation (I) process models (Andersen, 1993).

The U-model (Johanson and Vahlne, 1977) is based on the behavioural theory of the firm (Aharoni, 1966; Cyert and March, 1963) and the growth of the firm (Penrose, 1959). The model argues that a firm bases its foreign-market entry and market country selection on its current stock of experiential knowledge. Firms accumulate experiential knowledge by operating in the international market. Decisions are made as problems or opportunities arise. When faced with a decision in the international market, firms apply those solutions that have been successfully applied in the past. Experiential knowledge reduces the risks of operating in the international market. Objective knowledge, on the other hand, is obtained through activities such as market research etc. Thus, internationalization is a matter of learning, about foreign markets, foreign clients, cultures and institutions, as well as internal resources and capabilities of the firm (Eriksson *et al.*, 1997). The model postulates an incremental interplay between market knowledge and market commitment.

The U-model states that internationalization is an uncertainty reducing incremental process, starting in countries at a short psychic distance[1] from the domestic market. Psychic distance is defined as factors preventing or disturbing the flow of knowledge and experience to and from the market (Hörnell, Vahlne and Wiedersheim-Paul, 1982; Nordström and Vahlne, 1992). The model views foreign-market entry as beginning with occasional export orders, followed by regular exports, the establishment of local sales subsidiaries and, subsequently, local manufacturing. Collecting experience on clients abroad needs interacting with them. But not all service firms require an equal amount of interaction with their counterparts (in terms of number, frequency and intensity) to accumulate experiential knowledge. The experiential knowledge of a firm influences the selection of foreign market as well as the choice of foreign-market entry mode.

In the initial years of foreign-market entry the experiential knowledge that the firm has gained from its domestic market is of limited value in markets located at great psychic distance. Firms with little experience of foreign markets prefer markets at a short psychic distance. Among Swedish firms the first foreign market was Norway or Denmark. Investigating US firms, Davidson (1980, 1982, 1983), detected a preference for English-speaking countries, like Canada and Australia. Vernon (1966) observed a steady shift away from culturally familiar to culturally less familiar markets. He found that firms with a vast stock of experience show less preference for markets at a short psychic distance.

Researchers have reported parallel findings in studies of service firms, such as banks (Khoury, 1979) and advertising agencies. Weinstein (1974) notes that in their early foreign market activities, US advertising firms prefer culturally similar markets. Hisrich and Peters (1985) reported that US banks perceived less uncertainty about those Eastern European markets which they had experience. Sharma and Johanson (1987) find no evidence that technical consultancy firms first enter markets at a small psychic distance. Erramilli (1991) and Erramilli and Rao (1993) found that US-based service firms start their foreign-market entry in countries at a short psychic distance.

There are indications in the literature that firm's size may condition choice of foreign markets. While studying the international experience of US firms, Erramilli (1991) found that as service firms grow larger they increasingly enter culturally distant markets. Empirical research by Eriksson *et al.* (1997) found no correlation between size and international experience.

Foreign market involvement and knowledge

According to the U-model the lack of knowledge about foreign markets is the main obstacle to internationalization and knowledge can mainly be developed through experience from operations in those markets. Foreign business opportunities and problems are discovered through experiences from foreign markets operations. Experience gives the firm an ability to see and evaluate business opportunities and thereby to reduce uncertainty associated with commitments to foreign markets. Since knowledge is developed gradually, international expansion takes place incrementally. But, as Figure 12.1 shows, entry modes differ from each other based on resource commitment and integration/control

Entry mode	Form	Integration/Control	Commitment
1. Wholly-owned subsidiary	Subsidiary	high	high
2. Partly-owned subsidiary	Minority/majority ownership, affiliaties, etc.	high/moderate	high/moderate
3. Contract, alliances	Relationship	moderate	low
4. Market	Exports	low	low

Figure 12.1 Characteristics of firms entering foreign markets
Source: Based on Majgård and Sharma (1998).

(Anderson and Gatignon, 1986). FDI modes, such as wholly-owned subsidiaries, require more resource commitment than exporting and contractual transfer modes. They also involve higher levels of risk. Gatignon and Anderson (1988) observe an increasing propensity to select wholly-owned subsidiaries with increasing experiential knowledge. Also, prior experience of producing in a market is related to a firm's preference for wholly-owned subsidiaries. Firms resort to joint ventures and licensing to a lesser extent in markets at small psychic distances. The use of licensing and joint ventures increases in markets that are dissimilar to the domestic market. Kogut and Singh (1988) found that US firms move from ownership to non-ownership-based foreign-market entry modes as they move away from culturally similar markets. This is due to the fact that a firm's existing experiential knowledge is less relevant in culturally dissimilar environments. The absence of experiential knowledge results in faulty estimates of the cost of foreign-market entry and of the revenues, as well as creating uncertainty regarding market demand. Firms that are uncertain of demand for their products and services in a country market opt for low resource commitment (Calof and Beamish, 1995; Gatignon and Anderson, 1988). Firms can also form alliances (Dunning, 1995) or enter into franchise agreements abroad.

As stated earlier, evidence from the service sector is limited. Foreign market selection of most service firms may be clustered into three main areas, such as exporting, licensing and FDI. Some services can be produced in the seller's home country and transported (via disks) to buyers in the foreign country (e.g., computer software). The use of external agents and distributors is not uncommon for these services, although direct-to-customer export channels appear to be the preferred modes. Services such as management consulting may be less exportable since they require service production mainly at the client's location. Non-equity forms of investment such as franchising, licensing or management contracts are more dominant in services such as fast food, hotels, car rentals and retailing. Knowledge-based and people-embodied services are not engaged through licensing due to the danger of lower performance of their services.

Client-following and market-seeking strategies

Erramilli and Rao (1990) and Majkgård and Sharma (1998) identified two distinctive foreign-market entry strategies: client-following and market-seeking. The latter is an offensive approach and refers to when a firm enters foreign markets primarily to serve customers abroad. Client-following is defensive, and means that the firm follows its existing domestic clients, suppliers, cooperation partners etc. abroad, i.e., those firms are already part of an international business network. The results indicate that market-seeking firms are likely to employ entry modes that involve cooperating partners, for example, agents, exports, joint ventures and contractual agreements.

The explanation for the above mentioned phenomena lies in the nature of the knowledge and network of relationships of firms. We define business networks as sets of interconnected business relationships, in which each exchange relation is between business firms conceptualized as collective actors (Anderson, Håkansson and Johanson, 1995). This research demonstrated that close, lasting relationships between firms doing business with each other are critically important for knowledge accumulation in firms. It takes time and resources to build relationships. According to this research all firms are engaged in a limited set of business relationships with customer and supplier firms, which, in turn, have relationships with still other firms. Commitments are made to specific business firms whether they are customer firms, supplier firms, intermediary firms or other cooperating firms. They will also primarily concern gradual development of the relationships in which the firm is already engaged. In a similar way as posited by the internationalization process literature the business firms will develop close interdependencies in relation to the important partners and they will consequently be prepared to defend those relationships by increasing the commitments to those firms with which they already do business. They will also tend to develop other relationships which can be expected to support the important relationships. This research shows that networks are a source of knowledge for internationalizing firms.

Client-following firms show less preference for low integration and control entry modes. This provides further support to the notion that in committing resources to foreign markets decision-makers are more influenced by experiential knowledge. Erramilli (1991) investigated the impact of experiential knowledge in foreign market selection. Two dimensions of experiential knowledge were identified: length, the number of years the firm was engaged in operations abroad prior to the current entry, and variation of this experience. The results confirm that experiential knowledge plays an important role in explaining the internationalization process of firms. The results suggest that firms choose similar foreign markets at low level of experiential knowledge, and unfamiliar countries at a higher level of experience. In market selection, the geographic variation of a firm's experience was more influential than its length of experience. Firms with little experiential knowledge appear to show high affinity for full-control entry modes. The findings suggest that a high degree of control in firms is possible with a limited degree of resource commitment.

Internationalization of hard and soft services

Regan (1963) and Rathmell (1974) state that services are intangible, heterogeneous, and perishable, and that the production and consumption of services are inseparable. Services are a product for which personal contacts between the buyer and seller are important. Heterogeneity poses a problem of attaining consistency and quality control in delivery of the services to the client. Finally, services cannot be stored. Knowledge and trust are therefore important in services, as is the reliability and availability. This makes some services experience something whose value to the buyer becomes apparent only after use (Nelson, 1970).

Erramilli (1990)[2] divided services into hard services and soft services. Hard services (e.g. insurance, software services, engineering and architecture) permit separation of production and consumption. In soft services (education, management consulting, etc.) it is often difficult to separate production from consumption and that is one reason why export of soft services is difficult (Carman and Langeard, 1980). Consequently, hard services and soft services differ in their knowledge accumulation and internationalization process. Hard services internationalize presumably in the same way as manufacturing industries. Soft services differ.

Services with a relatively high degree of tangibility and a low demand for physical interaction between producer and customer can be exported, or licensed (Sampson and Snape, 1985; Vandermerwe and Chadwick, 1989). For example, architectural services can be transferred in a document, a diskette or other tangible medium. Today it is even possible to export software development services and accountancy services via links and satellites from India to the USA. Soft services require local proximity of service providers and service buyers.

The data collection and sample

A mail survey of Swedish firms engaged in international operations was conducted to collect data for this study. A sample of these service firms was drawn from three business directories (the Swedish Trade Registers, service industry branch registers and the Confederation of Swedish Enterprise). 1206 service companies, representing a variety of service industries, were included in the mail survey. Questionnaires were mailed to CEOs and presidents most likely to be involved in the foreign-market entry decision process in these firms. Each respondent provided data on the first foreign-market entry decision. We received 958 usable responses. The response rate is 74.5 per cent.

All of the questions are of close-ended nature using a five-point Likert scale, ranging from 'not at all important' to 'very important'. Most of the variables are measured as a function of the perceptions of the managers. These perceptions may vary according to variations in their past experience of international operations, level of knowledge about a country, etc. Sample characteristics are summarized in Table 12.1. There is a broad representation of sectors (e.g. management consulting, advertising, computer software and data processing, education, accounting, legal, engineering and architecture firms), and there is a good distribution of firms in relation to size and year of first foreign-market entry.

Several studies in international business incorporate size as an important independent variable (Bonaccorsi, 1992; Calof, 1994). It is suggested that the size of a firm correlates positively to the development of knowledge. The sample spans service firms with one employee to those with several thousand employees. More than half of the sample is made up of firms with 99 or fewer employees. These service firms may be considered large in their particular sector but are small in comparison to manufacturing firms that go abroad.

Hard services and soft services is a dichotomous variable. Examples of hard services are computer software and data processing, architecture, and miscellaneous engineering services. Soft services include legal,

Table 12.1 Characteristics of foreign-market entries in sample ($N = 958$)

	% Firms
A. Distribution of firms: by size	
No. of employees	
a. 1–49	43.3
b. 50–99	12.4
c. 100–499	21.7
d. 500–10000	22.6
B. Distribution of firms: year of first foreign-market entry	
Year of first foreign-market entry	
a. 1895–1949	11.2
b. 1950–1969	13.4
c. 1970–1979	24.7
d. 1980–1989	19.5
e. 1990–2002	31.2
C. Distribution of firms: by industry	
Service industry	
Management consulting and Education	23.6
Advertising	17.5
Computer software and Data processing	14.1
Legal and Accounting	17.5
Engineering and Architecture	17.5
Miscellaneous services	9.8
D. Distribution of firms: by type of service	
Type of service	
1. Hard service	45.2
2. Soft service	54.8
F. Distribution of firms: by entry mode	
Entry mode	
1. Wholly-owned high-control entry mode	58.2
2. Low-control entry mode (includes contractual method, cooperation, joint venture, exports)	41.8

advertising, accounting, education, banks, and management consulting. The distribution between hard and soft service firms is fairly balanced with 433 (45.2 per cent) hard service firms, and 525 (54.8 per cent) soft service firms.

102 questionnaires were later dropped from the analysis for having missing values on specific variables. The remaining 958 questionnaires supplied good quality data with few missing values. In some cases (53 firms out of 958), the service firm had planned their first foreign assignment but had not yet entered the foreign market. These are included in the sample.

In order to establish whether the differences between respondents and non-respondents affected the results, a non-respondent analysis was done. The sample was split, on the basis of the survey return date, into early and late respondents. An analysis, in which late respondents are viewed as non-respondents (Armstrong and Overton, 1977), shows that non-respondents did not differ significantly from respondents in terms of size and service industry sector. The results may be regarded as representative for the non-respondents as well.

Empirical findings

Tables 12.1–12.6 show some basic characteristics of the data. The variables represent descriptive raw data on the reasons for going abroad, mode of entry, geographical area of first foreign-market entry and foreign market strategy, first foreign-market entry and psychic distance and first foreign-market entry mode and type of service. The response rates vary due to missing values (see separate notes for each table).

According to the internationalization process model, the duration of international operations affects the development of experiential knowledge. It is postulated that firms that have been abroad long have more experiential knowledge than firms which have recently gone abroad. In the sample, some firms have international experience of one year, and others have been abroad for more than 50 (one firm had been operating abroad for 108 years). Table 12.1 shows that most of the firms in the sample made their first foreign-market entry in the 1990s. Few of the firms started their foreign operations before or during the 1950s. 37 firms have not yet carried out an assignment abroad, but intend to do so in the future.

It is natural to ask about the reasons for going abroad. As stated earlier, in going abroad firms are either market seekers or client followers. Some firms are already involved in international business relationships abroad. Other firms must create a position in this network of relationships. As seen in Table 12.2, there are a number of different reasons for going abroad. The most important of these is the firm's desire to gain new knowledge. The indicator 'higher profits abroad' was an inadequate reason for going abroad. However, no single reason was sufficient to explain why firms went abroad.

International cooperation between firms has attracted growing attention. Several researchers stress that cooperation has increased dramatically in recent years and that more firms enter international markets via cooperation now than ever before (Contractor and Lorange, 1988;

Parkhe, 1993). It is therefore interesting to investigate further how common international cooperation is with firms in comparison to firms that went abroad alone as market-seeker. Table 12.3 shows two different foreign market strategies, namely market-seeker and client-follower. As stated earlier, these two types of internationalization strategies demand two different types of knowledge about international markets and clients. The result shows that the most common foreign-market entry mode is market-seeker (44.5 per cent) in comparison to firms that went abroad in cooperation with others, client-followers (36.2 per cent). The number of missing values is 184. Client-followers use their existing network of relationships to gain experiential knowledge of the foreign market, which results in less uncertainty. Firms going abroad alone need to invest more resources, for example, to identify customer requirements.

Table 12.2 Reasons for going abroad

Reasons for going abroad	*Mean score*	*No response*	*Total*
To gain new knowledge	4.1	96	958
To establish closer relationship with clients	3.7	88	958
To serve home base clients worldwide	3.6	127	958
Demand from established clients	3.5	92	958
Higher profits abroad	3.2	103	958
To provide and maintain services' quality	3.1	78	958
To expand the market	2.9	132	958
To find new markets	2.8	176	958
To go abroad when competitors did so	2.6	133	958
To exploit the size of the market	2.2	165	958
To gain bigger market share	1.9	179	958
To gain direct market access	1.9	184	958

Managers were requested to rate their answers on a five-point Likert scale. Mean score of categories (1 = not important and 5 = very important).

Table 12.3 Mode of entry

Mode of entry	*n*	*%*
Market-seeker (alone)	426	44.5
Client-follower (member in a consortium/cooperation/subcontractor to Swedish or foreign firm etc.)	346	36.2
No response	184	19.3
Total	958	100

Further analysis shows that firms going abroad for the first time more commonly go abroad alone than in cooperation with others. In their first foreign-market entry, most early and late starters have gone international alone rather than in cooperation with others. The results also show that going abroad alone has increased in the late starter group in comparison with early starters. One reason could be that Swedish economy is highly international. Even small and medium-size Swedish firms are engaged in exchange with internationally active firms. This network of relationship is a source of knowledge about foreign markets and clients. Table 12.4 shows that the most common strategy of first foreign-market entry in Scandinavia is market-seeking strategy, and the least common is to carry out foreign assignments as a member in an international network, as a subcontractor to foreign consultants in the same service sector, i.e., client-following strategy. The most common foreign market strategy in the rest of the world is client-following. The latter use their existing network of business relationships to gain knowledge of the foreign market and clients, which results in less uncertainty foreign-market entries may also follow when a firm's partner demands that the firm accompanies them abroad or otherwise is interested in extending their relationship to encompass business also in foreign markets. This is frequently the case when the firm enjoys close relationships with firms that are internationalizing (Majkgård and Sharma, 1998). In this case, the foreign-market entries are likely to occur in those countries where the focal firm's customers are operating rather than in countries with large markets or countries on a short psychic or cultural distance. Another consequence of client-following is that the firm's internationalization may be quite rapid. As stated earlier, these two internationalization strategies require different types and amounts of knowledge on foreign markets and foreign clients.

Table 12.4 Geographical area of first foreign-market entry and foreign-market strategy

Foreign-market strategy	*Geographical area of first foreign-market entry*	
	Scandinavia	*Rest of the world*
Market-seeker	488 (51%)	423 (44%)
Client-follower	470 (49%)	535 (56%)
	958 (100%)	958 (100%)

The only way of learning about an international network is to start interacting with one or several of the network actors and thereby learn how strong and connected their relationships are. This means that the outsiders cannot comprehend foreign country market networks, even if they are aware that the networks exist. Each actor only knows about its own specific network context and has vague ideas about the more distant network patterns, whether this concerns technologies, product markets, country markets, or any field.

Table 12.4 shows that existing business relationships can be used in accumulating knowledge and climbing over the country market barriers and entering more culturally distant markets. The network view assumes that there may be several ports of entry and, consequently, several ways of getting into the country market. The focal firm may try to establish a relationship with a customer firm in a foreign market. When doing so there are several different firms to approach and since each potential counterpart is unique and may require a specific knowledge and relationship development process and also can be expected to have its particular consequences, the decision to approach one is an important and difficult decision. Internationalizing firms apply the method of trial and error. The establishment and development of knowledge and relationship is a result of a mutual interaction. Thus, development of knowledge in a foreign market is a complex, uncertain and time-consuming process, which may require considerable commitment of time and other tangible resources on the part of the internationalizing firm.

While there is a qualitative difference between foreign-market entry and foreign market expansion in the earlier internationalization models, we argued that they concern similar problems. Foreign market expansion is a matter first of developing the firm's relationships in the specific market, second of establishing and developing supporting relationships, third of developing relationships that are similar to, or connected to the focal one. Although all this development may be confined to one country market it may also cross country borders and lead to entry into other foreign markets. As our case shows, in order to support a strategic relationship abroad the firm may be forced to develop a relationship in another country, thereby entering that country market.

The existing literature on foreign-market entry suggests that firms start internationalization in culturally similar countries and then incrementally move into more unfamiliar countries. Table 12.5 presents a profile of the first markets entered abroad by firms in our sample and a comparison with the psychic distance index by Hörnell *et al.* (1973) which are based on manufacturing industries. The data indicates that

Table 12.5 First foreign-market entry and psychic distance

First foreign-market entry[a]	*Psychic distance*[b]
Norway	Denmark
Denmark	Norway
Germany	Germany
England	Finland
USA	England
Finland	USA
Holland	France
France	Japan
Belgium	Holland
Japan	Switzerland
Australia	Spain
Switzerland	Belgium
Italy	Austria
Argentina	Italy
Canada	Canada
South Africa	Brazil
Spain	Portugal
Ireland	South Africa
Portugal	Argentina
Brazil	Australia

[a] n = 958
[b] Psychic distance index based on Hörnell, Vahlne and Wiedersheim-Paul (1982).

the first foreign-market entry of Swedish firms was Scandinavia or Western Europe. Nearly half of all first foreign entries are made in Norway, followed by Denmark, Germany and England. Worth noting is that there is practically no difference between the ranking of countries of first entry and the ranking of countries based on the psychic distance. The correlation between the two indices is high. We, thus, conclude that the fundamental process of going abroad is the same in service industries as in manufacturing industries. Table 12.6 shows that high-control entry modes being clearer regarding soft services (64 per cent) than hard services (39 per cent). Hard services use more low-control foreign entry modes (61 per cent) in comparison to soft services (36 per cent).

One explanation is that hard services with a relatively high degree of tangibility and low demand for physical interaction between producer and customer can be exported or licensed. Today it is even possible to export software development services and accountancy services via links and satellites from, for example, India to the US. Conversely, soft

Table 12.6 First foreign-market entry mode and type of service

First foreign-entry mode	*Type of service*	
	Hard services	*Soft services*
High control	374 (39%)	613 (64%)
Low control	584 (61%) 958 (100%)	345 (36%) 958 (100%)

services where production and consumption occur simultaneously are difficult to export (Erramilli, 1990). Presumably, based on the viability of decoupling, hard service industries internationalize in the same manner as manufacturing industries, while internationalization of soft service industries is unique. Given the simultaneous production and consumption of a soft service, the service provider will be reluctant to relinquish control in uncertain circumstances, thus in foreign markets soft service firms are more likely to opt for high-control entry modes than hard service firms. Therefore, soft service firms are more likely to choose a high-control entry mode than hard service firms. We conclude the internationalization process of firms within service industries may differ considerably.

The central assumption concerning development of learning and foreign experiential knowledge in the Uppsala process model corresponds to the empirical analysis presented in the paper, but the process manifests itself in a different form than that seen in the manufacturing sector. The analysis shows that the internationalization process of hard services is identical to those of manufacturing firms. These firms go abroad slowly.

Implications for knowledge management

In this chapter we have outlined a knowledge-based model of the internationalization of the firm. We identified the experiential learning-commitment interplay as the driving focusing on business network relationships. In the empirical findings we can see firms learning in relationships, which enable them to enter new markets abroad in which they can develop new relationships, which in turn give them a platform for entering other foreign markets. These have implications for knowledge management in internationalizing firms.

Our findings indicate that the world being structured in national entities with different cultures and institutional settings matters less than some have earlier believed. The process of globalization implies countries over time become more similar to each other in terms of culture and institutional settings. And firms develop their knowledge base to manage national institutions and national clients. We believe that there has been an overemphasis on psychic distance between countries in explanations of the internationalization process. Maybe individual relationships can change the perception of psychic distance thanks to experiential learning. Similarly, the entry mode issue, which is so important in much writing on internationalization, is less important in our analysis. By stressing the significance of knowledge and lasting relationships between autonomous actors we reduce the importance of ownership control and thereby the whole ownership issue, which has had such a dominating role in international business. This too has implications for knowledge management in firms.

The process of knowledge accumulation and utilization in the internationalization process in service industries seems to be almost identical to the processes observed in manufacturing industries. In service industry firms, as in manufacturing firms, knowledge is accumulated based on trial and error. As service firms operate in international market, they learn from clients and institutions. This is a time- and resources-consuming process. There are significant costs involved in collecting and using knowledge. In this connection, it is important for firms to identify and be aware of their existing internal knowledge management routines, and associated constraints. Service firms need to develop a cognitive framework to identify, collect, interpret and transform experience into knowledge. Developing structures and routines to facilitate learning is a fundamental process for a firm's international competitiveness. Development of experiential knowledge requires repetition and variation in a given set of activities, for example, development of foreign business relationships in the foreign-market entry. The study shows that, as in case of manufacturing firms, experiential knowledge influences the internationalization of service firms. These are developed in relationships with particular customers and suppliers. The findings point out the importance of network relationships in the internationalization process of firms. The differences in knowledge management in service firms and manufacturing firms are more of degree and less of kind. Moreover, our results support the findings by Sharma and Johanson (1987), Erramilli and Rao (1993) and Eriksson *et al.* (1997), showing that the fundamental process of internationalization in manufacturing and service industries is more or less identical. Moreover, no special FDI-MNE theories for international service firms are necessary (Buckley, Pass and Prescott, 1992; Boddewyn, Halbrich and Perry, 1986).

We detected significant differences within service industries. We argue that differences in knowledge management in firms seems to be associated more with (1) the nature of the service (that is, hard *vs* soft service), and the internationalization strategy of firms (client-following *vs* market-seeking strategy).

First, as argued earlier, hard and soft service firms require different amounts and types of knowledge about clients and markets in going abroad. Our results show that soft service firms in their internationalization process are more likely to choose a high-control entry mode than hard service firms. In order to manage relationships with international buyers, the routines in soft service firms need to be capable of collecting more knowledge on buyer, their needs and their decision-making processes. Soft service firms also need to accumulate more knowledge, as well as more differentiated knowledge, on institutions, rules and regulations and resources abroad. These firms need more sophisticated and multi-level knowledge collection and knowledge interpretation routines than hard service firms. Management of knowledge in soft service firms seems to be more demanding and resources-consuming than in hard services.

Second, we earlier stated that client-following and market-seeking firms are engaged in different types of network of relationships, and they possess different types of knowledge. To client-following firms the existing business network relations offer strong opportunities for international expansion. Client-following firms are, thus, engaged in a significant amount of pre-foreign-market entry learning. They have developed knowledge accumulation and knowledge management routines in interaction with as well as adapted to specific clients. These firms have already committed customer-specific resources to facilitating exchange with specific clients. The client-specific experiential knowledge for internationalization already exists at the head office in the firm. The issue of knowledge management in client-following firms is, thus, mainly a matter of (1) the codification of an existing knowledge and (2) selecting either one or a combination of appropriate media to 'transport' this existing stock of tested knowledge to overseas markets. In the process of going abroad, the challenge client-following firms face is how to transfer knowledge abroad, but without introducing major changes in their current stock of knowledge. Moreover, the direction of knowledge flow is one way, from the head office to foreign markets.

In other words, the knowledge management practices and processes in client-following firms are more exploitation-oriented.

Market-seeking firms are different and these firms internationalize in order to seek new business partners abroad. Market-seeking firms face the challenge of developing new relationships and network of exchange abroad. They lack experience of the international markets and of specific foreign clients, and must acquire experiential knowledge of the international market and international clients. Domestic-based knowledge accumulation and knowledge interpretation routines and practices may not be suitable abroad. The internationalizing firm may need either new knowledge management routines or the existing knowledge management routines in the firm may require major modifications. The market-seeking firms are, thus, exposed to two challenges. First, these firms may be required to unlearn some of the already institutionalized knowledge and business practices in the firm. A simple transfer of domestic-based business routines abroad may harm the firm. Secondly, knowledge management in market-seeking firms is associated with accumulating and interpreting new knowledge on (potential) clients, suppliers, competitors and institutions abroad. On the surface it might seem that it is easy to network, just as if climbing a ladder from relationship to relationship. But this is wrong. The building of business network relationships abroad is a complex and delicate matter, which requires resources and time as well as responsiveness to the interests of the specific partners. This is a gradual process involving significant risk and cost. The flow of knowledge is between the head office of the firms and the foreign markets. Learning and integrating new knowledge about clients, markets and competitors is critical. But, the head office in the firm must also supply knowledge to the units located abroad. The knowledge management processes in market-seeking firms may be more exploration-oriented, based on trial and error. An extreme case is firms that are international from the time of their establishment (Oviatt and McDougall, 1994; McDougall, Shane and Oviatt, 1994). In the initial year of internationalization these firms lack knowledge management routines and network of relationships. They need to evolve internal knowledge management routines, in parallel with developing network of exchange with clients and suppliers in international markets.

Suggestions for future research

This study makes a contribution to the ongoing research on the internationalization process of firms in general and service firms in particular. Various related problems await study in the future. First, there are reasons to investigate the antecedents of experiential knowledge in the internationalization process. What characterizes the experiences that lead to development of experiential knowledge? In the literature, it is generally assumed that experience is closely related to the length of time of foreign business operations (Barkema *et al.*, 1996). Erramilli (1991) suggests that variation of experience is relevant to knowledge development. Consequently, it can be expected that variations in countries, in modes of operations and cooperative partners, are important antecedents of experiential knowledge. In internationalization studies, it is also assumed that learning about foreign markets requires a high degree of local presence, which may also be considered antecedent to experiential knowledge.

For knowledge management in firms it would be relevant to investigate to what extent experiential knowledge is country-specific or general. Is it transferable from one subsidiary to another in the same firm, or across national boundaries? Several researchers (Johanson and Vahlne, 1977; Chang, 1995) have suggested that experiential knowledge is not transferable. Later research has indicated possibilities of experience transfer (Makino and Delios, 1996). There is a need to investigate the conditions for transfer of experiential knowledge between different company units.

Studies of internationalization have indicated that two different kinds of market-specific experience can be distinguished, business experience and institutional experience (Eriksson *et al.*, 1997). Business experience concerns experience related to the business environment of the firm, which according to the business network view comprises the firms with which it is doing business or is trying to do business with. Institutional experience concerns such factors as language, laws, regulations and public and semi-public authorities implementing laws and regulations. The distinction between those two kinds of experience means that we have reason to expect that they are developed and managed in different ways. More research is needed.

Notes

1. Gatignon and Anderson (1988) use the term socio-cultural distance, and Kogut and Singh (1988) the term cultural distance to denote the same or very similar phenomena. Nordström and Vahlne (1992) state that the phenomena of psychic distance and cultural distance overlap. In this chapter, the terms psychic distance and cultural distance are used interchangeably.
2. See also Sampson and Snape (1985) and Boddewyn *et al.* (1986).

References

Aharoni, Y. (1966) The foreign investment decision process. *Boston: Division of Research, Graduate School of Business Administration, Harvard University.*

Andersen, O. (1993) On the internationalization process of firms: a critical analysis. *Journal of International Business Studies*, 24(2): 209–32.

Anderson, E. and Gatignon, H. (1986) Modes of foreign entry: a transaction cost analysis and propositions. *Journal of International Business Studies*, 17(3): 1–26.

Anderson, J. C., Håkansson, H. and Johanson, J. (1995) Dyadic business relationships within a network context. *Journal of Marketing*, 58(October): 1–15.

Armstrong, J. S. and Overton, T. S. (1977) Estimating non-response bias in mail surveys. *Journal of Marketing Research*, 14(3): 396–402.

Axelsson, B. and Johanson, J. (1992) foreign-market entry: the textbook *vs* the network view. In B. Axelsson and G. Easton (eds) *Industrial Networks: A New Reality*. London: Routledge, pp. 218–34.

Barkema, H., John, H., Bell, J. and Penning, J. M. (1996) Foreign entry, cultural barriers, and learning. *Strategic Management Journal*, 17(2): 151–66.

Benito, G. and Gripsrud, G. (1992) The expansion of foreign direct investments: discrete rational location choices or a cultural learning process? *Journal of International Business Studies*, 23(3): 461–77.

Boddewyn, J. J., Halbrich, M. B. and Perry, A. C. (1986) Service multinationals: conceptualization, measurement and theory. *Journal of International Business Studies*, 17(3): 41–57.

Bonaccorsi, A. (1992) On the relationship between firms size and export intensity. *Journal of International Business Studies*, 23(4): 605–35.

Buckley, P. J., Pass, C. L. and Prescott, K. (1992) The internationalization of service firms: a comparison with the manufacturing sector. *Scandinavian International Business Review*, 1(1): 39–56.

Calof, L. J. (1994) The relationship between firm size and export behavior revisited. *Journal of International Business Studies*, 2: 367–87.

Calof, L. J. and Beamish, P. W. (1995) Adapting to foreign markets: explaining internationalization. *International Business Review*, 4(2): 115–31.

Chang, S. J. (1995) International expansion strategy of Japanese firms: capability building through sequential entry. *Academy of Management Journal*, 38(29): 383–407.

Contractor, F. and Lorange, P. (1988) Cooperative strategies in international business. Lexington, MA: Lexington Books.

Cyert, R. M. and March, J. (1963) A behavioural theory of the firm. New York: Prentice-Hall.

Davidson, W. H. (1980) The location of foreign direct investment activity: country characteristics and experience effects. *Journal of International Business Studies*, 11(2): 9–22.

Davidson, W. H. (1982) *Global Strategic Management*. New York: John Wiley.

Davidson, W. H. (1983) Market similarity and market selection: implications of international marketing strategy. *Journal of Business Research*, 11: 439–56.

Dunning, J. H. (1995) Reappraising the eclectic paradigm in an age of alliance capitalism. *Journal of International Business Studies*, 26(3): 461–92.

Eriksson, K., Johanson, J., Majkgård, A. and Sharma, D. (1997) Experiential knowledge and cost in the internationalization process. *Journal of International Business Studies*, 28(2): 337–60.

Erramilli, M. K. (1990) Entry mode choice in service industries. *International Marketing Review*, 7(5): 50–62.

Erramilli, M. K. (1991) The experience factor in foreign-market entry behaviour of service firms. *Journal of International Business Studies*, 22(3): 479–501.

Erramilli, M. K. and Rao, C. P. (1990) Choice of foreign-market entry mode by service firms: role of market knowledge. *Management International Review*, 30(2): 135–50.

Erramilli, M. K. and Rao, C. P. (1993) Service firms international entry mode choice: a modified transaction-cost analysis approach. *Journal of Marketing*, 57: 19–38.

Gatignon, H. and Anderson, E. (1988) The multinational corporation's degree of control over foreign subsidiaries: an empirical test of a transaction cost explanation. *Journal of Law, Economics and Organization*, 4(2): 305–35.

Hisrich, R. D. and Peters, M. P. (1985) East–west trade: an assessment by US banks. *Columbia Journal of World Business*, 20(1): 15–22.

Hofstede, G. (1980) Culture's consequences: international differences in work-related values. Beverly Hills, CA: Sage.

Hörnell, E., Vahlne, J.-E. and Wiedersheim-Paul, F. (1982) *Export och utlandsetableringer...* Stockholm: Almqvist & Wiksell.

Johanson, J. and Wiedersheim-Paul, F. (1975) The internationalization of the firm: four Swedish cases. *Journal of Management Studies*, 12(3): 305–22.

Johanson, J. and Vahlne, J.-E. (1977) The internationalization process of the firm – a model of knowledge development and increasing foreign market commitments. *Journal of International Business Studies*, 8(1): 23–32.

Khoury, S. J. (1979) International banking: a special look at foreign banks in the US. *Journal of International Business Studies*, 10(3): 36–52.

Kogut, B. and Singh, H. (1988) The effect of national culture on the choice of entry mode. *Journal of International Business Studies*, 19(3): 411–32.

Majkgård, A. and Sharma, D. D. (1998) Client-following and market-seeking strategies in the internationalization of service firms. *Journal of Business-to-Business Marketing*, 4(3): 1–41.

Makino, S. and Delios, A. (1996) Local knowledge transfer and performance: implications for alliance formation in Asia. *Journal of International Business Studies*, 27(5): 905–28.

McDougall, P. P., Shane, S. and Oviatt, B. M. (1994) Explaining the formation of international new ventures: the limits of theories from international business research. *Journal of Business Venturing*, 9: 469–87.

Nelson, P. (1970) Advertising as information. *Journal of Political Economy* (July–August): 729–54.

Nordström, K. A. (1991) *The Internationalization Process of the Firm: Searching for New Pattern and Explanations*. Ph.D dissertation, Stockholm School of Economics, Institute of International Business.

Nordström, K. A. and Vahlne, J. E. (1992) Is the globe shrinking? Psychic distance and the establishment of Swedish sales subsidiaries during the last 100 years. Paper presented at the International Trade and Finance Association's Annual Conference, April 22–25, Laredo, Texas.

Oviatt, B. M. and McDougall, P. P. (1994) Toward a theory of international new ventures. *Journal of International Business Studies*, 25(1): 45–64.

Parkhe, A. (1993) Strategic alliance structuring: a game theoretic and transaction cost examination of interfirm cooperation. *Academy of Management Journal*, 36: 794–829.

Penrose, E. (1959) *The Theory of the Growth of the Firm*. Oxford: Basil Blackwell.

Petersen, B. and Pedersen, T. (1997) Twenty years after – support and critique of the Uppsala internationalization model. In I. Björkman and M. Forsgren (eds) *The Nature of the International Firm*. Copenhagen: Copenhagen Business School Press.

Rathmell, J. M. (1974) *Marketing in the Service Sector*. Cambridge, MA: Winthrop.

Regan, W. J. (1963) The service revolution. *Journal of Marketing*, 47(July): 57–62.

Sampson, G. P. and Snape, R. H. (1985) Identifying the issues of trade in services. *World Economy*, 8 (June): 171–82.

Sharma, D. and Johanson, J. (1987) Technical consultancy in internationalisation. *International Marketing Review*, 4(Winter): 20–9.

Stopford, J. M. and Wells, L. (1972) *Managing the Multinational Enterprise*. New York: Basic Books.

Vandermerwe, S. and Chadwick, M. (1989) The internationalization of services. *Service Industries Journal*, 9(1): 71–93.

Vernon, R. (1966) International investment and international trade in the product cycle. *Quarterly Journal of Economics*, 80(2): 190–207.

Weinstein, A. K. (1974) The international expansion of US multinationals advertising agencies. *MSU Business Topics*, 22(Summer): 29–35.

Zejan, M. C. (1990) New ventures or acquisitions: the choice of Swedish multinational enterprises. *Journal of Industrial Economics*, 38(3): 349–55.

13
Plumbing and Plugging-In: Networking by Venture Capitalists in Europe and the USA

Anna Gatti and Morten Thanning Vendelø

Introduction

The new economy represents a world marked by uncertainty and change. A key driver of the new economy is information technology, which fuels the new economy with new applications to be both explored and exploited. Firms operating in this environment face 'hypercompetition' (D'Aveni, 1994), and they must constantly acquire and try out new combinations of knowledge if they are to survive. Hence, in this environment fast creation and acquisition of new knowledge are important to firms.

The venture capitalist industry is emerging as a global industry within the new economy, as venture capitalists invest in start-ups located in different countries. Furthermore, the venture capitalist industry is global as technologies developed and introduced to the market in one part of the world are likely to influence the profitability of venture capitalist investments in other parts of the world. One important consequence of the globalization of this industry is that investment decisions by venture capitalists are fairly complex and information intensive. They are information intensive in the sense that much information is of relevance to decisions, and complex as many aspects of this information need to be taken into consideration when making investment decisions. Furthermore, the environment in which venture capitalists operate is one of uncertainty and change, as the firms involved constantly discover and define new potentials for the information technology. In such an environment knowledge is quickly up- and out-dated. Consequently, when facing investment decisions venture capitalists face high search costs for information. It has been suggested that organizations operating under such conditions cannot search and learn fast enough on their own to obtain a satisfactory level of knowledge and information. Thus, access to other sources of information and knowledge through networks is crucial (Powell *et al.*, 1996: 116):

> when the knowledge base of an industry is both complex and expanding and the sources of expertise are widely dispersed, the locus of innovation will be found in networks of learning, rather than in individual firms.

As a result we must expect that when operating in hypercompetitive environments firms with extensive networks will perform better, as they have access to more knowledge, and thus have a better chance of survival. Adding to this argument, Tsai (2001: 996) suggests that organizations occupying network positions with access to new knowledge developed by many other units produce more innovations and enjoy better performance. Hence, it seems that the closer you get, the better access to knowledge you have, and thus, the better off you are.

Our study of networking by venture capitalists challenges this assumption, as our observations suggest that, in hypercompetitive environments, networks that allow for quick connections and acquisition of knowledge in terms of opinions and information are very useful for decision-making. We call these *plug-in networking processes*. Plug-in networking processes seem inexpensive to develop, maintain and use, as well as valuable to venture capitalists facing decisions such as which technologies to invest in and which to exit from. However, we know little about the characteristics of plug-in networking processes, how these processes are developed, maintained and used, and thus there is reason to take a closer look at them. We do so by drawing on an explorative study of networking by a European venture capitalist, which was found to draw extensively on networks in order to inform its investment decisions.

The research questions addressed are:

> How are plug-in networks established and maintained? (The plumbing question)
> When is knowledge transferred in plug-in networks?
> What kind of knowledge is transferred in plug-in networks? (The plug-in question)

The chapter proceeds in the following manner. First, we discuss knowledge flows in an international business perspective, thereby situating our research questions in this context. Second, we describe three branches of network research. Third, we compare network behaviour and investment patterns by US and European venture capitalists. Fourth, we describe the methods used in our study of networking by a venture capitalist. Fifth, we report the findings from this study, and finally, we discuss the implications of and perspectives provided by our findings.

Knowledge flows in an international business perspective

In recent years networking and knowledge flows have increasingly attracted interest from researchers working in the field of international business. Foremost, this interest is propelled by the observation that rapid technological change and globalization of industries are important trends in the environment inhabited by a growing number of companies. As noted by Ariño and de la Torre (1998: 306):

> The complexity of organizational tasks required by technological acceleration and the rapid globalization of markets have made it increasingly difficult for any one firm to go at it alone in all product/markets of interest. Thus, inter-firms collaboration is 'a major topic of interest and relevance in the present organizational world' (Smith *et al.*, 1995: 20).

Two branches of research exist, both focusing on information and knowledge flows in an international context. In international business research a common theme is strategic alliances and the evolution of these. Within this branch one finds a subset focusing on joint learning and knowledge flows. For example, Larsson *et al.* (1998) look at alliances as dynamic processes with the regard to processual barriers to joint learning, and Kumar and Nti (1998) model the evolutionary process of knowledge-intensive learning alliances. As a central foundation for their work Kumar and Nti use the concept of absorptive capacities (Cohen and Levinthal, 1990) to describe the abilities of firms to benefit from such alliances. Emphasizing the importance of absorptive capacity Andersson *et al.* (2001: 8) suggest that the distance between the knowledge bases of two entities involved in a knowledge flow determines the opportunities for actual transfer of knowledge.

> The longer the gap between the accumulated knowledge in the transferring and the recipient firm the more difficult and less effective the transfer. If the gap between the knowledge base of the two firms is

large then the recipient will fail to catch and decode the incoming signals from the knowledge transferring firm.

In the other branch of research on alliances and networking by multinational corporations (MNC) we find a perspective which emphasizes long-term and intense collaboration, thereby implying that firms operate in fairly stable environments where dependence on a limited number of sources of knowledge is relevant. Most studies of networking by MNCs focus on firms as actors in networks. For example, Andersson *et al.* (2001: 2) note: 'Through the social network, the firm can get access to resources and capabilities from outside the organization, such as capital goods, services, innovations, etc.'

Pedersen *et al.* (2001: 1) emphasize that effective dissemination throughout the MNC organization of knowledge acquired by its local affiliates is an important source of competitive advantage. Thus, it is essential that the MNC applies a transfer mechanism that suits the specific knowledge characteristics. Hence networking by MNCs is conceptualized as relationships between organizational entities, but does not include a micro-level understanding of how these relationships are established. This also implies that we have a different view on how networks contribute to the competitive advantage of firms. Within the MNC literature it is common to describe the network benefits as 'capabilities' or 'resources'. We want to be more specific and refer to the benefits as 'decision information and knowledge'. We choose this as the venture capitalist industry is one of decisions. Venture capitalists constantly face decisions about entrance to, continuation of and exit from investments. Such decisions involve considerations regarding the prospects of new technologies, e.g., new IT applications, the future of markets. Much information on these issues of course flows from business plans, progress reports, etc., delivered by entrepreneurs and start-up firms. Yet, this information might be coloured and sugar-coated, possibly leading to decisions that in the long run turn out to be bad decisions. Other sources of decision information and knowledge are therefore needed in this uncertain world, as noted by Audretsch and Thurik (2001: 290–1):

> The expected value of a new idea, or potential innovation, is likely to be anything but unanimous between the inventor of that idea and the decision maker, or group of decision makers, of the firm confronted with evaluating proposed changes or innovations.

We analyse a different way of acquiring information for decisions. Our research builds on network literature, as it is commonly accepted that networks are important sources of knowledge both within and among organizations (Granovetter, 1973; Hansen, 1999; Powell *et al.*, 1996).

Three streams of network research

Networks facilitate exchange, sharing and transfer of knowledge among individuals and organizations. Social network analysis seeks explanations for social and economic phenomena, starting from the web of social relations in which individuals engage, rather than from their psychological or demographic attributes (Wellman, 1988). Social network analysis significantly developed in the 1970s with the seminal work of Harrison White (e.g., White *et al.*, 1976). Since then network analysis has contributed significantly to our understanding of social and economic life (Nohria and Eccles, 1992).

Three streams of research in network analysis are considered here. The first conceptualizes social networks as conduits of information and ideas, and has been developed in many different areas of sociological and organizational inquiry (Burt, 1987). We call this information networks. *Information networks* are particularly important for recognizing opportunities. The second stream of research stresses networks' ability to signal the actors' status to every player in the game (Podolny, 1993). Since status and reputation are critical in accessing financial capital and other types of resources, *status networks* are very important for venture capitalists and entrepreneurs. The third tradition of research focuses on the ability of certain types of network structure to engender trust, facilitating or inhibiting collective action (Coleman, 1988). We call these mobilization networks. *Mobilization networks* are critical when entrepreneurs try to put together resources from different sources, and coordinate the start-up effort.

The social networks literature provides a useful starting point for understanding how social relations affect entrepreneurship as, in the context of entrepreneurial action, social networks can be conceptualized as providing the critical connections that allow entrepreneurs and venture capitalists to recognize opportunities, acquire resources, mobilize interest, and coordinate collective action successfully (Aldrich and Zimmer, 1986). Yet, for the most part, social network research focuses on the structural properties of the networks of social relations and does not provide any guidance on the strategies and practices for managing social relations (Salancik, 1995). A possible explanation for this gap in the social network research agenda in entrepreneurship stems from the lack of ethnographic knowledge about the entrepreneurial action. In order to move beyond a purely structural network perspective, this chapter suggests a research strategy to develop a relational approach to entrepreneurship that emphasizes the ability of venture capitalists to build, shape and use their networks. This research strategy shifts the focus of the investigation from the networks to *networking*, which can be defined as 'the set of practices and strategies actors employ to create, maintain, use, and make sense of their networks of social relations' (Ferraro, 2001).

Venture capitalism in the USA and in Europe

The American venture capitalist industry has been booming in the past decade. This entrepreneurial form, developed in the Silicon Valley in the 1970s, has been recognized as one of the most important facilitators of the new economy. Most studies of networking by venture capitalists focus on the US venture capitalist industry (Brophy, 1996) and the Silicon Valley in particular (Saxenian, 1996). In 1986 there were approximately 150 independent venture capital firms in the United States (Azouaou and Magnaval, 1986; Davis, 1986), including more than 140 with a presence in Silicon Valley alone. Venture capitalists 'perform a gatekeeping function, intervening to help create new companies and actualize important breakthroughs' (Florida and Kenney, 1988: 128), ensuring a liaison function between various sources of funds and new companies (Azouaou and Magnaval, 1986). A comparison of Silicon Valley to European industrial network districts such as northern Italy reveals that both regions tend to rely on a dense network of supporting institutions: universities, venture capital firms, consulting groups (Harrison, 1994). Furthermore, both regions rely on local production relationships, although Silicon Valley's relationships extend to multinational firms (Harrison, 1994). However, while being organized around networks, Silicon Valley businesses rarely enter long-term contracts with each other, and tend to be fiercely competitive, showing no hesitation to drop a relationship if a better deal comes along (Harrison, 1994), hence relying on a hybrid of network and market arrangements, leading to 'cooperative–competitive' relationship patterns. Venture capital firms are important actors in this network, acting not only as source of capital but as 'system integrators, bringing otherwise independent firms together to join forces in making some new product or pursuing some new process development' (Harrison, 1994: 109). As a network of organizations, Silicon Valley is characterized by intense informal socialization and information exchange between companies (Saxenian, 1996). This pattern of collaboration–competition is characteristic of the entire high-tech sector of Silicon Valley. While collaborating with global industries, Silicon Valley firms tend to collaborate with local companies for most production relationships (Harrison, 1994). The region of Silicon Valley is hence based on a flexible network form of organizing, in which companies alternatively collaborate and compete against each other (Saxenian, 1996): 'from the outside, the industry appeared to be small, tightly knit, quite homogeneous in strategy and practice, and unusually cooperative in doing deals... Venture capitalists shared deal flow and information concerning due diligence and co-invested frequently' (Bygrave and Timmons, 1992: 27), and top venture capitalists in Silicon Valley form a highly tight-knit network (Bygrave and Timmons, 1992).

In the American new economy, the venture capital firm is a clearly institutionalized type of organization with formalized rules and standards, codified behaviour and role (Suchman, 1995). In Europe the diffusion of this kind of organization started about ten years ago. Thus, the diffusion of the 'start-up phenomenon' in Europe has a characteristic of novelty, and the 'new economy' is 'new' only for the European region, but is already 'old' (in term of institutionalized rules) in the United States. Although young compared to the United States industry (OECD, 1996), the venture capitalist industry in Europe is assuming a significant role in the market. Cambridge University has taken the initiative in strengthening university–industry cooperation (*Economist*, 1999), following a similar European trend (Owen-Smith *et al.*, 2002). Furthermore, although the bulk of the European venture capitalist funds go to low-tech companies, investment in high-technology companies is on the rise (PricewaterhouseCoopers, 1999).

Investment patterns in the USA and in Europe

The geography of Silicon Valley, consisting of a narrow stretch of land between the bay and the foothills, ensured dense geographical development, fostering more ties within the region (Saxenian, 1996). The tendency for venture capital firms to concentrate in a specific region seems to be the result of a need for uncertainty-reduction through increased information-exchange. Venture capitalists are typically reluctant to invest in remote companies, leading to 'a natural agglomeration among the universities, spin-off firms, and ventures', characteristic of venture capitalist industries in California, Massachusetts, and the London area (Leinbach and Amrhein, 1987: 150). Indeed, more than 70 per cent of Californian venture capitalist investments are located within the state of California (Schilit, 1990). An important study by Bygrave and Timmons (1992) analysed network patterns of co-investing in American venture capital from 1967 through 1982, comparing venture capitalists in the high-tech versus the low-tech sectors. Their results revealed that venture capitalists operating in high-tech co-invest significantly more than venture capitalists operating in low-tech. Network density was significantly higher for high-tech venture capitalists, and co-investment was higher for early stage ventures. Certain key characteristics associated with geography can partly explain the differences between Silicon Valley and the United Kingdom. The most important one is the formation of an industry around institutions of higher education. Silicon Valley relies on strong ties to Stanford University (Bygrave and Timmons, 1992) and other regional universities. By contrast, the venture capitalist industry in Italy as well as in other European countries generally lacks a well-developed university–industry cooperation infrastructure (Vander Weyer, 1995).

The classic American definition of venture capital pertains to the funding of seed and early-stage ventures (Bygrave and Timmons, 1992). By contrast, European venture capitalists embrace a different definition: they tend to fund mature industries, and define early-stage companies as up to three years old, an age which would no longer be considered a start-up in the United States (Bygrave and Timmons, 1992). The Italian venture capitalist industry reflects the European trend of investing less in early-stage companies than their US counterpart (OECD, 1996). In the Italian venture capitalist industry it is also common to invest less in new technology. Italian venture capitalists prefer investing in a process-based business idea than in a product-based one (Grandori, 2001). This different business model corresponds to a different venture capitalist profile: more technical and specialized in the Silicon Valley and more managerial in Italy (Grandori, 2001). This implies that Italian venture capitalists have good experience in managing organizational resources, processes and politics, but lack technical knowledge to judge innovative technologies.

Another key characteristic of the venture capitalist industry in Silicon Valley is the hands-on approach of the investors. Venture capitalists in

Silicon Valley are mostly former entrepreneurs (many from the region's high-tech industry) (Saxenian, 1996), and become actively involved in the creation and the management of start-ups (Florida and Kenney, 1988; OECD, 1996). In exchange for funding, Silicon Valley venture capitalists ask for a main stake in ownership of the company, a say in the selection of the start-up's management, and ongoing advice and assistance (Bygrave and Timmons, 1992; Davis, 1986; Florida and Kenney, 1988). Armed with such expertise and the potential very high returns, Silicon Valley venture capitalists are willing to take risks with their investments. As entrepreneurs themselves, venture capitalists value learning (Banatao and Fong, 2000; Vander Weyer, 1995) and are sometimes willing to work with an entrepreneur with a history of failure (Banatao and Fong, 2000). A fundamental culture of trust is cultivated between venture capitalists and investment targets (Davis, 1986). By contrast, venture capitalists in Italy lack start-up experience as entrepreneurs and do not actively participate in the companies that they fund (Grandori, 2001). Their industry is characterized by risk aversion and investments in more mature companies or management buyouts (Vander Weyer, 1995). The process of finding venture capital money for start-ups is thus much longer in Europe than in the USA, and failure is seen as a liability (Vander Weyer, 1995). In sum, this tells us that European venture capitalists face 'problems' in obtaining qualified knowledge and information needed to evaluate potential investments.

In an earlier study (Gatti and Vendelø, 2001) we found some evidence of a common behaviour among European venture capitalists. Behind active relationships between companies (partnership, investments, ownerships and joint venture), it seems that European venture capitalists rely on advice by experts outside their formal network when trying to evaluate business opportunities. These are not necessarily involved in formal links, as tracked by relationships, but they orbit in some way around venture capitalist networks. They are, for example, engineers employed in some hot start-up, whom venture capitalists chanced to meet at a convention or social event. These persons do not belong to venture capitalists' formal and stable network, but they have the competence that venture capitalists are looking for at that specific time.

Methodology

Using the three streams of network research as inspiration, we studied how decision-makers in European venture capital firms search for and obtain information needed for their investment decisions. We looked at the networks developed by an Italian venture capitalist in order to understand which kind of networks are activated and what dynamics are generated. In order to gather the empirical data needed to investigate cross-national networking by venture capitalists, we designed an exploratory case study (Yin, 1984) of networking by an Italian venture capitalist. In Italy, as in other European countries, the diffusion of venture capitalists is relatively recent. We identified one of the most representative venture capital firms in Italy, and gathered information on the perspectives of two levels of the management hierarchy in the firm. In order to identify our case study, we have taken into account the media coverage and we have defined 'representative venture capitalists' as actors in the same field defined it. That is to say, before selecting our representative firm, we made investigative inquiries in the field, asking people to suggest the three most representative venture capital firms in Italy, and we selected our case study from those suggested.

We collected data through interviews, observations, and secondary sources. The primary source was semi-structured interviews with individual respondents. At the site we interviewed founders, CEO and COO of the venture capital firm. We conducted interviews during a several-day site visit. Interviews typically lasted 60 minutes, although a few ran as long as two hours. During the site visit we kept a record of impressions and recorded informal observations that we made as we participated in such activities as lunches, quarterly board dinners, coffee breaks, company soccer games and product demonstrations. In addition, whenever possible, we attended meetings as passive note-takers. These observations provided real-time data.

We used two interview guides to conduct two levels of semi-structured interviews. In both cases, we asked respondents open-ended questions that let them tell their stories of how they decided to explore new investment opportunities and validate existing investments, and how they acquired information to judge high-tech business plans. We asked probing questions to establish details (i.e. when a particular event occurred). Moreover, in order to analyse critical competence and how it was acquired, we used the behavioural event interview (BEI). This methodology aims to identify on which competence a positive or negative solution of a critical event was based. Interviewees were asked to recall a critical decision-making event related to high-tech investments when they felt they had been in control of the situation, and which they could class as a successful decision-making process. Then, they were asked to focus on competence, skills and information sources that were crucial to the decision. Similarly, the interviewee was asked to recall an event where s/he felt uncomfortable, and which s/he could describe as a non-successful decision-making process. Then, s/he was asked to focus on a competence and information source that was critical in that situation. In this way, we collected information on the decision-making process and the information-collection process by the Italian venture capitalist. The interviews were taped, transcribed and analysed using ATLAS.it software. First, we entered all transcribed responses into a database indexed by interview number, interview type, and question number. Next, we did text analysis using ATLAS.it to identify logical linkage and correlation between interviews and concepts. Using these interviews and secondary sources (documents, records, etc.), we completed the case study. As a check on the emerging case story, we asked a third researcher to read the original interviews and form an independent view on it (Eisenhardt, 1989).

Results

The analysis of our empirical data shows that venture capitalists operate in a business environment where time is perceived to be an extremely scarce resource. Many decisions are to be made regarding investments in start-ups. High-profiled venture capitalists receive a large number of applications for funding from entrepreneurs, and thus many opportunities occur and need to be evaluated. The large number of applications implies a huge variation in the technological approaches taken by the entrepreneurs, meaning that the venture capitalist commonly does not have the knowledge needed to evaluate the applications. In addition, the background and educational profiles of European venture capitalists (financial and marketing backgrounds dominate) often put them on shaky ground when judging the technological approaches described in the applications. Hence, there is much incentive to seek information and knowledge elsewhere. But acquiring knowledge is often time-consuming and costly, and time and money are not exactly slack resources in the new economy. In particular the decision-making process regarding new investments has to be fast, because it is important to arrive before competitors.

In our study we observed flows of information and not real flows of knowledge in the plug-in networks. Both levels of interviewees from the venture capital firms describe patterns of information search where they make use of links to people whom they might only have met once, and briefly. Such encounters might have taken place at business meetings, while travelling in foreign countries, or at occasional dinner parties, and the link may not have been either maintained or used since then. Yet, when information needs appear, links are (re)enacted, using either e-mail or telephone, in order to obtain the specific information needed, and then they dematerialize again. We interpret this as plug-in behaviour, and suggest that it can be conceptualized as weak–weak ties. There seems to be no reciprocity involved, meaning that there is no expectation about payback later on, which thus may or may not happen. Hence, we think that there is reason to suggest that network behaviours by Italian venture capitalists are less network-structure related and more networking-process related.

Our observations concur with Granovetter's (1973) observation that people to whom one is weakly tied are likely to move in different circles, are thus less likely to be connected to one another. These contacts are likely to be relatively more heterogeneous in information than contacts to whom one is strongly tied, as the latter are more likely to be tied to one another. The more complex the knowledge to be transferred the higher the tie 'bandwidth' needs to be (Hansen, 1999), and thus for the transfer of complex knowledge strong ties are better than weak ones. If the network transfers knowledge, the final result will be that the recipient can rely on her/his acquired knowledge to take decision in future similar situation. In the plug-in situation, the Italian venture capitalist acquires the necessary information to take a specific decision, without even investigating the knowledge behind the information acquired. This means that in the next similar situation the recipient will be unable to take a decision autonomously and s/he will plug-in again, to the same source of information or to another. Plugging-in the same source of information is not common, as venture capitalists prefer to change their source of information even if the kind of information they need is the same. This has two reasons behind it. First, the Italian venture capitalist prefers to diversify the risk and to maintain multiple information sources. For example, if the venture capitalist decides not to invest in a technology-based business plan based on the information obtained by plugging-into source A, s/he will hardly know in the short run if s/he received the right advice from that source. Thus plugging-in a different source of information next time minimizes the risk of relying on an unreliable information source. As Cohen and Levinthal note, 'If there is uncertainty about which domain will be the source of useful new information, heterogeneous background of knowledge helps increasing the likelihood of relatedness, and thus of learning effectiveness.' To keep multiple information sources active, the networking process is crucial and critical. Second, using the same source of information several times generates an indebtedness which the venture capitalist might wish to avoid.

Discussion

Our results regarding plug-in networking by the Italian venture capitalist suggest that we need to revise our understanding of networks. In particular we need to ask: What is the network conception behind the plug-in process? Why is plugging-in an alternative to other types of networking? We do so by discussing alternative conceptions of networks. First, we discuss networks as cognitive constructs, thereby, focusing on both the formation and maintenance of plug-in ties. Second, we discuss trust in plug-in networks, foremost focusing on why venture capitalist can rely on the information obtained in plug-in situations.

Networks as cognitive constructs

Based on the results from our exploratory study, it makes sense to conceptualize plug-in networks as cognitive constructs, which enables people to rationalize in the following way: 'I remember that we had a pleasant meeting sometime in the past, and therefore, I can approach you for advice.' Hence, plumbing happens and ties exist because individuals form and hold personal narratives about past interactions (including their emotional associations), feeling good or bad about them. About this phenomenon White (1992: 67) notes:

> A tie becomes constituted with story, which defines a social time by its narrative ties. A social network is a network of meanings as Burns & Burns emphasized long ago. (1973: 16–18)

Hence, our perspective is not entirely new, as White (1992: 65) suggested that network ties can be thought of as stories: 'Networks are phenomenological realities as well as measurement constructs. Stories describe the ties in network.' According to White (1992), each network tie is the residue of some event or sequence of events involving two related actors, and Pentland *et al.* (2002: 7) elaborate by emphasizing that 'Between individuals, these events may involve communication, workflow, or other activities ... The story may be very simple and short (a one-time, arms-length purchase of a standard item), or complicated and long (collaborative design and fabrication of a new kind of electronic device).' Yet, here we go beyond the interpretation of networks as stories provided by Pentland *et al.* (2002), as we argue that stories are used by actors to conceptualize and re-enact ties with other actors. Again as noted by White (1992: 84): 'Stories are vital to maintaining as well as generating social spaces for continuing actions.'

Ergo, ties come into existence as actors retell stories of past interactions. However, as in Pentland *et al.* (2002), we maintain that 'a tie represents a relational event, or a collection of relational events. Thus, we can think of a plug-in network as a summary of the relational events in some collection of stories. If we collect a sample of stories about helping or advice, we could observe an advice network. If we collect manufacturing stories, we could observe a supplier network. The particular network that we see depends on what kinds of stories (and relational events) we choose to emphasize, and the time period over which we collect our observations' (Pentland *et al.*, 2002: 14–15).

In the case of plug-in networks, these exist as a large collection of short stories, each of which may only rely on few encounters with the information source. Namely, the initial event where the venture capitalist came to know the potential source of information, and the few subsequent incidents where the venture capitalist plugged into the source of information. The point being made here is that networks exist as and are maintained in stories of narrative quality. We thus subscribe to the idea that human cognition has a narrative structure, implying that networks are cognitive constructs. Our conception of plug-in networks challenges the idea that frequency of interaction strengthens networks. In our view usefulness of ties is not a matter of frequency and durations of interactions, instead, it is a matter of perception.

Swift trust in plug-in networks

Our research also challenges conventional conceptions of trust in networks. It is common to think about trust as evolving from either institutions or repeated interaction between parties over a period of time. When evolving from repeated interaction over a longer period of time trust serves to eliminate opportunistic behaviour, and thereby it makes future interactions appear more secure. As noted by Jarvenpaa and Leidner (1999: 809): 'The traditional conceptualization of trust assumes that trust resides in personal relationships and past or future memberships in common social networks that define the shared norms of obligation and responsibility.'

What we have identified in the case of the Italian venture capitalist seems to be incidents of swift trust, which allow actors to rely on information provided by parties with whom they only had brief interaction sometime in the past. The concept of swift trust was developed by Meyerson *et al.* (1996) to describe trust in teams formed around a common task with a finite life span. These temporary teams consist of members with diverse skills, a limited history of working together, and little prospect of working together again in the future. 'The tight deadlines under which these teams work leave little time for relationship building. Because time pressure hinders the ability of team members to develop expectations of others based on firsthand information, members import expectations of trust from other settings with which they are familiar' (Jarvenpaa and Leidner, 1999: 794). Thus, members of such teams rely on the initial use of category-driven information processing to form stereotypical impressions of others. This implies that 'unless one trusts quickly, one may never trust at all' (Meyerson *et al.*, 1996: 192); that is, it is the initial social exchange of a relationship that facilitates trust (Jarvenpaa and Leidner, 1999: 806). In contrast to the swift trust described by Meyerson *et al.* (1996: 180) the swift, trust employed by venture capitalists is not strengthened through a 'highly active, proactive, enthusiastic, generative style of action'. Instead, our findings concur with those of Jarvenpaa and Leidner (1999: 794), who note:

> whereas traditional conceptualizations of trust are based strongly on interpersonal relationships, swift trust de-emphasizes the interpersonal dimensions and is based initially on broad categorial social structures and later on action. Because members initially import trust rather than develop trust, trust might attain its zenith at the project's inception. (Meyerson *et al.*, 1996)

Again it is the scarcity of time that forces actors to attribute trust immediately, because there is no alternative. Going back to the conception of networks as cognitive constructs, we suggest that trust is based on the emotions attached to the first interaction, thus trust is a matter of whether the actor feels good or bad about this first encounter with a potential source of information. In the specific setting that we studied, another reason for the venture capitalist to trust is that its business is not in competition with the provider of information, and thus no conflict of interest is involved.

Closure

This study focuses on how a relatively young organization, a venture capital firm in Italy, supports decisions such as which technologies to invest in and which to exit from, in a not yet established institutional settings. As pointed out above, the venture capitalist industry in Europe is still young compared to the United States venture capitalist industry, and is still identifying the best practices and institutionalized rules of behaviour. In this context, our findings show that the Italian venture capitalist has developed a multi-node map of sources that are activated only in order to retrieve the necessary information to judge technical business plans. This means that the venture capitalist has a clear map of who knows what. Each point of this map is not, as in the well-structured long-term network, a node linked to the organization on a regular basis, but it is generally a person who was occasionally met in a conference or during a social event. When the information need appears, the link is re-enacted in order to obtain the specific information needed. It can be conceptualized as a weak–weak tie. We have defined this behaviour as a plug-in process. This process starts by drawing the new source of information in the organizational map (plumbing) and thereafter it is poorly maintained through occasionally telephone calls or social meetings. The organization does not really invest to maintain it, whereas it invests to expand its map of sources.

The key issue here is not to understand the structure of the network, but to understand the process of networking. We argue that in unstable, time-constrained, multi-national and high-tech environment, venture capitalists rely more on a dynamic process of networking than on well-structured network based on long-term relationships, also in order to avoid a real mutual interdependency. This argument contradicts conventional wisdom in international business research, which emphasizes the importance of long-term and intense collaboration for knowledge to flow. We show that under certain circumstances short-term relationships are preferable, even in an international context.

Our findings have some interesting implications. First, more studies should focus their attention on the process of networking instead of the structure of networks. Since in this kind of dynamic and unstable entrepreneurial environment the critical step is the effective plug-in and the transfer of information, the structural analysis of networks will give less and less critical explanation of the situation.

Second, we appreciate that, also in this study, institutional settings have a determinant impact on organizational behaviour (DiMaggio and Powell, 1983). In the Silicon Valley, where venture capitalists act in a more mature and dense industry, the traditional analysis of network structure can offer useful and meaningful explanations (Bygrave and Timmons, 1992; Harrison 1994). In Italy, where the venture capitalist industry is young and counts few actors, the process of networking is more explicative of how venture capitalists take decisions on high-tech business plans. However, another feature is crucial in order to understand the reason why plug-in networking processes emerged as the way of acquiring critical information to support technical decisions: the education and background experience of venture capitalists. In a comparative study between Italian and Silicon Valley venture capitalists (Grandori, 2001), it appears that in Silicon Valley the technical background (i.e., engineers, computer scientists) dominates while in Italy the financial and marketing background characterizes venture capitalists. This obviously implies knowledge constraints in judging high-tech business plans. Developing a networking process facilitates information collection and reduces the risk of establishing a strong tie with an unreliable sources of information. In fact, on the one hand, technical competences of Italian venture capitalists are not strong enough to screen reliable sources of high-tech competence; on the other hand, the pace of the industry does not allow for waiting to see if the advice given by one node was good or not. High-tech business plans, presenting fūtūristic technologies, arrive on the venture capitalist's table one after the other, and decisions need to be taken ahead of the competition. In this industry, following means losing. Plugging-in different nodes, activating diverse competence sources, can reduce the risk of relying on the wrong counterpart.

Third, we argue that, while in the Silicon Valley networks there is a knowledge transfer (Hansen, 1999; Saxenian, 1996), in the plug-in networking there is only transfer of information. This means that, in the first case, the recipient will be able to rely on his/her acquired knowledge to take a similar decision in the future; in the second case, the recipient will be unable to take autonomously a decision in a future similar situation, but s/he will plug into her/his network to acquire once again the necessary information, without even decoding the knowledge behind it. Thus, while in the well-structured networks we can talk about knowledge transfer, in the plug-in networking we can only talk about information transfer. Moreover, we can argue that Silicon Valley networks can be identified as mobilization networks, whereas plug-in networking can be interpreted as the process behind information networks.

Finally, we should assume a different point of analysis for discussing the plug-in networking process. In our study, we have analysed the problem from an internal point of view. This means that we have discussed the plug-in networking processes and networks as source of information (or competence, in the case of plug-in behaviour) from the point of view of the actors involved in. However, networks *per se* can be information for an external observer. They can tell you which alliance you will likely do in the future months, in which technology you are going to invest, from which centre of excellence you are draining brains, and thus on which competences you will rely. In a sentence, studying network structure can be an excellent source of competitive intelligence. If, instead of building long-term structured networks, the organization develops dynamic processes of plug-in networking, it becomes more difficult for competitors to acquire critical information by studying the structure of network. In fact, having the 'picture' of the activated network at the time T1 does not mean much, since it will be only the portrait of the enacted plug-in at that time. This information will be of poor help to a competitor at the time T2. Thus, we also argue that promoting plug-in networking processes do not only allow them to rely on multiple sources of competence, but can also reduce the risk of competitive intelligence activities.

References

Aldrich, H. and Zimmer, C. (1986) Entrepreneurship through social network. In D. Sexton and R. Smilor (eds) *The Art and Science of Entrepreneurship*. New York: Ballinger, 3–23.

Anderson, U., Forsgren, M. and Holm, U. (2001) The strategic impact of relative embeddedness in external networks: subsidiary performance and competence development in the multinational corporation. LINK Conference, September 2001: 1–45.

Ariño, A. and de la Torre, J. (1998) Learning from failure: towards an evolutionary model of collaborative ventures. *Organization Science*, 9: 306–25.

Audretsch, D. B. and Thurik, A. R. (2001) What's new about the new economy? Sources of growth in the managed and entrepreneurial economics. *Industrial Corporation and Change*, 10: 267–315.

Azouaou, A. and Magnaval, R. (1986) *La Silicon Valley, un marché aux puces*. Paris: Ramsay.

Banatao and Fong (2000) The Valley of Deals: how venture capital helped shape the region. In C. M. Lee, W. F. Miller, M. G. Hancock, and H. S. Rowen (eds) *The Silicon Valley Edge: A Habitat for Innovation and Entrepreneurship*. Stanford, CA: Stanford University Press.

Brophy, D. (1996) United States venture capital markets: changes and challenges. In OECD (ed.) *Venture Capital and Innovation*. Paris: OECD, Working Paper No. 98.

Burns, T. and Burns, E. (eds) (1973) *The Sociology of Literature and Drama*. Harmondsworth: Penguin Books.

Burt, R. (1987) Social contagion and innovation: cohesion versus structural equivalence. *American Journal of Sociology*, 92: 1287–335.

Bygrave, W. D. and Timmons, J. A. (1992) *Venture Capital at the Crossroads*. Boston, MA: Harvard Business School Press.

Cohen, W. M. and Levinthal, D. A. (1990) Absorptive capacity: a new perspective on learning and innovation. *Administrative Science Quarterly*, 25: 128–52.

Coleman, J. (1988) Social capital in the creation of human capital. *American Journal of Sociology*, 94: 95–120.

D'Aveni, R. (1994) *Hypercompetition*. New York, NY: Free Press.

Davis, B. M. (1986) Role of venture capital in the economic renaissance of an area. In R. D. Hisrich (ed.) *Entrepreneurship, Intrapreneurship, and Venture Capital*. Lexington, MA: Lexington Books.

DiMaggio, P. J. and Powell, W. W. (1983) The iron cage revisited: institutional isomorphism and collective rationality in organizational fields. *American Sociological Review*, 48: 147–60.

Economist (1999) Britain: ancient and modern. *The Economist,* 349: 84–5.

Eisenhardt, K. M. (1989) Building theories from case study research. *Academy of Management Review,* 14: 532–50.

Ferraro, F. (2001) Entrepreneurs and networking: towards a relational approach to entrepreneurship. Working paper presented at Department of Management Science and Engineering, Stanford University, Stanford, CA.

Florida, R. and Kenney, M. (1988) Venture capital and high technology entrepreneurship. *Journal of Business Venturing,* 3: 302–19.

Gatti, A. and Vendelø, M. T. (2001) Learning and adaptation by venture capitalists and start-ups in different cultural and institutional settings. Paper for the 17th EGOS Colloquium, Lyon, July 5–7. *Subtheme 12: European (ad)Venturing in the New Economy*: 1–30.

Grandori, A. (ed.) (2001) *Organizzazione e Governance del Capitale Umano*. Milan: EGEA.

Granovetter, M. (1973) The strengths of weak ties. *American Journal of Sociology,* 78: 1360–80.

Hansen, M. T. (1999) The search–transfer problem: the role of weak ties in sharing knowledge across organization subunits. *Administrative Science Quarterly,* 44: 82–111.

Harrison, B. (1994) *Lean and Mean: The Changing Landscape of Corporate Power in an Age of Flexibility*. New York, NY: Basic Books.

Jarvenpaa, S. L. and Leidner, D. E. (1999) Communication and trust in global virtual teams. *Organization Science,* 10: 791–815.

Kumar, R. and Nti, K. O. (1998) Differential learning and interaction in alliance dynamics: a process and outcome discrepancy model. *Organization Science,* 9: 356–67.

Larsson, R., Bengtsson, L., Henriksson, K. and Sparks, J. (1998) The interorganizational learning dilemma: collective knowledge development in strategic alliances. *Organization Science,* 9: 285–305.

Leinbach, T. and Amrhein, C. (1987) A geography of the venture capital industry. *Professional Geographer,* 40: 217–18.

Meyerson, D., Weick, K. E. and Kramer, R. M. (1996) Swift trust and temporary groups. In R. M. Kramer and T. R. Tyler (eds) *Trust in Organizations: Frontiers of Theory and Research*. Thousand Oaks, CA: Sage, pp. 166–95.

Nohria, N. and Eccles R. G. (eds) (1992) *Networks and Organizations: Structure, Form and Action*. Boston, MA: Harvard Business School Press.
OECD (1996) *Venture Capital and Innovation*. Paris: OECD.
Owen-Smith, J., Riccaboni, M., Pammoli, F. and Powell, W. W. (2002) A comparison of US and European university–industry relations in the life sciences. *Management Science*, 48: 24–43.
Pedersen, T., Petersen, B. and Sharma, D. (2001) Knowledge transfer performance of multinational companies. LINK Conference, September 2001: 1–36.
Pentland, B. T., Kwon, P. and Chung, M. J. (2002) Networks of action in distributed manufacturing: using process models to predict structural relationships. Paper presented at European Group of Organization Studies 18th Colloquium, Standing Workgroup 4: *Business Network Research*. Barcelona, July 4–6. 2002: 1–19.
Podolny, J. M. (1993) A status-based model of market competition. *American Journal of Sociology*, 98: 829–72.
Powell, W. W., Koput, K. W. and Smith-Doerr, L. (1996) Interorganizational collaboration and the locus of innovation: networks of learning in biotechnology. *Administrative Science Quarterly*, 41: 116–45.
PriceWaterhouseCoopers (1999) *Money for Growth: The European Technology Investment Report 1999*: EVCA.
Salancik, G. R. (1995) Wanted: a good network theory of organization. *Administrative Science Quarterly*, 40: 345–9.
Saxenian, A. (1996) *Regional Advantage: Culture and Competition in Silicon Valley and Route 128*. Cambridge, MA: Harvard University Press.
Schilit, W. K. (1990) *The Entrepreneur's Guide to Preparing a Winning Business Plan and Raising Venture Capital*. Englewood Cliffs, NJ: Prentice Hall.
Smith, K. G., Carroll, S. J. and Ashford, S. J. (1995) Intra- and inter-organizational cooperation: toward a research agenda. *Academy of Management Journal*, 38: 7–23.
Suchman, M. (1995) Managing legitimacy: strategic and institutional approaches. *Academy of Management Review*, 20: 571–610.
Tsai, W. (2001) Knowledge transfer in intraorganizational networks: effects of network position and absorptive capacity on business unit innovation and performance. *Academy of Management Journal*, 44: 996–1004.
Vander Weyer, M. (1995) The venture capital vacuum. *Management Today*, July: 60.
Wellman, B. (1988) Structural analysis: from method and metaphor to theory and substance. In B. Wellman and S. D. Berkowitz (eds) *Social Structures: A Network Approach*. Cambridge, UK: Cambridge University Press.
White, H. C. (1992) *Identity and Control: A Structural Theory of Social Action*. Princeton, NJ: Princeton University Press.
White, H. C., Boorman, S. A. and Breiger, R. L. (1976) Social structure from multiple networks: Part I. Blockmodels of roles and positions. *American Journal of Sociology*, 81: 730–80.
Yin, R. (1984) *Case Study Research: Design and Methods*. Beverly Hills, CA: Sage.

Index

Key: f = figure; n = note; t = table; **bold** = extended discussion or heading/word/phrase emphasized in main text.